Benjamin Franklin's Science

I. BERNARD COHEN

Benjamin Franklin's Science

HARVARD UNIVERSITY PRESS
CAMBRIDGE, MASSACHUSETTS
LONDON, ENGLAND
1990

Printed in the United States of America
10 9 8 7 6 5 4 3 2 1

This book is printed on acid-free paper, and its
binding materials have been chosen for strength and durability.

Library of Congress Cataloging-in-Publication Data

Cohen, I. Bernard, 1914–
Benjamin Franklin's science / I. Bernard Cohen.
p. cm.
Includes bibliographical references.
ISBN 0-674-06658-8
1. Franklin, Benjamin, 1706–1790—Knowledge—Physics.
2. Electricity—Experiments—History.
3. Lightning—Experiments—History.
4. Physicists—United States—Biography.
I. Title.
QC16.F68C64 1990
509.2—dc20 89-35290

Designed by Gwen Frankfeldt

For John Heilbron and Roderick Home,
historians of electricity in the age of Franklin
&
for Philip Morrison and
Gerard Piel and Dennis Flanagan,
three *Scientific American*s

Contents

Foreword

Benjamin Franklin is one of the best-known early Americans, yet his scientific work, which was fundamental and effective, is still not adequately understood. This book provides a full explanation of how his interest in science arose, the experiments he performed, and their influence throughout much of Europe as well as the United States. Although science was Franklin's primary concern during a relatively brief period of his life, it remained a fundamental part of his later associations and activities.

The full understanding of Franklin's actions throughout his lifetime requires the examination of his multiple roles. It has not been generally recognized that his work as a scientist has to be considered in order to evaluate his other activities. Through most of his life, Franklin's political leadership benefited significantly from his scientific reputation. His great success in supporting Pennsylvania and other colonies while he was in England, and later representing the new United States while in France, came in part because his scientific work was widely recognized. Of course, aside from politics, his scientific advances continued to be directly influential in furthering the understanding of electricity.

This study covers primarily the six years when Franklin was relatively free to study science, although I. B. Cohen also considers both his earlier works and the long balance of his life during which science remained significant to him.

Franklin began his career as a writer and printer; by the age of twenty-four he owned a printing business and a regularly published newspaper in Philadelphia. Through this work he became concerned with technical improvements in the city and began study and research in science, in which he became increasingly involved. In 1747 he withdrew from his printing business and until 1753 remained relatively free to carry out experiments and develop theories. His accomplishments in this period are explained here in well-written and understandable accounts.

Franklin was deeply influenced by Isaac Newton. He may have learned of his work when he first visited England. He followed not the *Principia,* Newton's most celebrated publication, but his *Opticks,* which was available in English. Its great advantage was presenting scientific studies through specific investigations and encouraging experiments.

The bulk of Franklin's scientific activities related to lightning and other electrical matters. His connection of lightning with electricity, through the famous experiment with a kite in a thunderstorm, was a significant advance of scientific knowledge. It found wide application in the construction of lightning rods to protect buildings in both the United States and Europe.

The story of the lightning experiments as well as analyses of other experiments are presented here. A recurring theme is the interaction of American and European developments. For example, Franklin's experiments with the Leyden jar, invented in the Netherlands, led to the development of a law of conservation of charge. In the other direction, Franklin's "sentry-box" lightning experiment was first successfully carried out in France.

I. B. Cohen has written extensively on Franklin's scientific writings and activities. In 1941 he published the first scholarly edition of Franklin's great book on electricity. Since then he has continued to contemplate Franklin's scientific history in numerous books and articles. The essays in this volume cover the entire range of Franklin's science.

It is important for the reader of these studies to keep in mind the meaning and importance of Franklin's work outside science. Franklin was a truly remarkable individual. These accounts clearly show his ability to discover new knowledge. They also reveal some of the ways in which his science added to his reputation and effectiveness in other fields. This volume demonstrates that Benjamin Franklin cannot be adequately evaluated without a clear grasp of the whole of his science.

Brooke Hindle

Preface

I have collected here some articles which I hope will prove to be of value to scholars interested in Franklin, in the history of physics, and in the development of science in America. Three chapters (1, 2, 5) have been written especially for this volume. The articles have been revised, but I have not attempted to provide references to all of the recent literature relating to Franklin and science. I have, however, introduced the results of the most significant additions to our knowledge with which I am familiar; I have also made corrections or emendations as needed, often as notes. In particular, I have taken advantage of the important researches of John L. Heilbron and Roderick W. Home. These are available in a number of articles and the following books: J. L. Heilbron, *Electricity in the Seventeenth and Eighteenth Centuries: A Study of Early Modern Physics* (Berkeley: University of California Press, 1979), and *Elements of Early Modern Physics* (Berkeley: University of California Press, 1982); Roderick W. Home, *The Effluvial Theory of Electricity* (New York: Arno Press, 1981); Roderick W. Home and P. J. Connor, *Aepinus's Essay on the Theory of Electricity and Magnetism* (Princeton: Princeton University Press, 1979).

When I first assembled this book, I showed it to Bern Dibner, a longtime admirer of Benjamin Franklin as a pioneer scientist. He pointed out that readers might be concerned by the lack of discussion of Franklin's most popularly known invention, the "Franklin stove." This omission was made all the more striking by the fact that I had devoted a considerable amount of space to a discussion of the relation between Franklin's science and his inventions, primarily in terms of the lightning rod. Furthermore, because I had devoted a chapter to researches made by Franklin in the domain of heat, the lack of reference to the stove was highlighted.

At this juncture, I learned that Sam Edgerton had written an essay on the history of the Franklin stove. I persuaded him to include in my book a revised version of his study. Thanks to his labors we now have a reliable account of Franklin's stove, presented in the context of inventions in heating technology with which Franklin was familiar. Edgerton also analyzes this invention in relation to Franklin's scientific thinking about heat.

One further remark needs to be made in relation to every selection reprinted here: my references to the earlier editions of Franklin's works (under the editorship of Sparks, Bigelow, and Smyth, respectively) should in every instance be updated by taking cognizance of the magnificent new edition, in the process of being completed, sponsored by the American Philosophical Society and Yale University and published by Yale University Press. This edition was begun under the general editorship of Leonard W. Labaree; later volumes have been edited by William B. Willcox. (Volume 26, covering the period from March 1 through June 3, 1778, was published in 1987.)

I am grateful to two students for their help in preparing this volume, Thomas Donovan and Ruth Oratz, and especially grateful to K. E. Duffin, who acted as editor and research assistant in all stages of the final redaction. I thank the editors and publishers of *Isis, Journal of the Franklin Institute, Pennsylvania Magazine of History and Biography,* and *Proceedings of the American Philosophical Society* for permission given to reprint articles originally published in those journals, and the Johns Hopkins University Press for permission to reprint my introduction to the book on the Pennsylvania Hospital.

Benjamin Franklin's Science

1

Introduction

Almost fifty years ago, when the first modern edition of *Benjamin Franklin's Experiments* was published, the presentation of Franklin's place in the history of science in general histories of American thought was characterized by ignorance and misunderstanding.[1] Franklin was portrayed in the popular role of gadgeteer and tinkerer; he was conceived by American historians to have been a clever inventor rather than a major scientist. His magisterial contribution to electrical science was usually reduced by American historians to the famous kite experiment and the invention of the lightning rod, and their true significance was not generally understood. One historian even went so far as to have his readers believe that Franklin "directed his scientific thought to the improvement of material conditions," as if Franklin could have predicted which aspect of his research might prove to be most fruitful for some unanticipated practical end![2] I very much doubt whether any American historian of the 1930s and 1940s would have been able to explain the sense in which Joseph Priestley could have legitimately compared Franklin to Newton.

In the 1930s, the major accounts in print of Franklin's scientific contributions were in German: portions of the histories of electricity by Ferd. Rosenberger and by Edmund Hoppe.[3] These were joined in 1937 by the extremely useful two-volume history of electricity, in Italian, by Mario Gliozzi.[4] In English, the chief sources of information concerning Franklin's contributions to electricity were the chapters in the popular histories of that subject by Brother Potamian (Michael F. O'Reilly) and James J. Walsh and by Park Benjamin.[5] Writers on Franklin still referred to a chapter on science in Paul Leicester Ford's volume of 1899, *The Many-Sided Franklin,* and a pair of short articles written in 1906 by Edwin J. Houston which treated Franklin's science primarily in relation to his mechanical inventions.[6] There was also a section of three pages in

Dorothy Turner's *Makers of Science,* written for a British secondary-school audience.[7] This literature, especially the works in English, was discursive and descriptive, and did not provide any real insight into Franklin's relation to the main themes in the development of scientific thought in the age of Newton.[8]

Among the reasons for the neglect of this aspect of Franklin's life and thought are two of major significance. One is that before World War II, pure or theoretical or fundamental or basic science did not yet occupy a high place in our national scale of values. Science was then largely supported by industry and America was still a long way from occupying a foremost scientific position as it does today.[9] Accordingly, the qualities Americans were wont to celebrate in a founding father of the republic might include thrift and industry, humor, business acumen, literary style, or even skill in diplomacy, but not theoretical science. Invention, yes; practical or applied science, yes; but pure or basic science, or fundamental research? Not really.

A second reason why Franklin's contribution to basic science was not part of the American historical tradition is that the history of science had not yet assumed its present place among the professional academic disciplines. In the 1930s and early 1940s, the history of science had not yet infiltrated the writings on (and teaching of) general history to the extent that it does now. Today's historians are all too aware of the historical force of science in our own times, and they quite naturally are alert to the place of the science of the past in the mainstreams of historical thought and actions. Hence, it would be no more than ordinary for a present-day historian to be genuinely curious about the significance of Franklin's science, whereas in the 1930s Franklin's scientific pursuits would have seemed only another index of his many-faceted mind or of his practicality. So young a subject is the history of science that in the mid-1930s there was not yet in the whole of the United States a graduate department offering a doctorate in this subject.[10] Today there are a good number. With an ever-growing body of professionally trained specialists in the history of science, and with an ever-increasing awareness of scientific issues on the part of general historians (both intellectual and social historians), it is only natural that at present there should be scholarly concern about the significance of Franklin's contribution to science in its many dimensions. Additionally, there has at last come into being a company of scholars whose primary area of study is science in America.[11]

The change that has occurred since the publication of *Benjamin Franklin's Experiments* can be seen by comparing Carl Van Doren's magnificent biography of Franklin to some recent additions to a continually increasing list. In 1938, Van Doren bunched together all of Franklin's fundamental contributions to science in one out of 26 chapters, a mere 27 pages in a 782-page full-scale account of Franklin's life and achievement.[12] With no fully adequate scholarly analyses to guide him, and without the training in science and its history necessary to make his own critical evaluation of Franklin's scientific contributions, Van Doren made use of the judgments by Franklin's contemporaries Joseph Priestley and John Pringle, whose evaluations he quoted at length. Contrast this with Thomas Fleming's *The Man who Dared the Lightning: A New Look at Benjamin Franklin* (1971), in which the opening pages are devoted to the theme of the title and to the electrical experiments that were the underpinning of the two lightning experiments he designed: the kite and the sentry box. Catherine Drinker Bowen's *The Most Dangerous Man: Scenes from the Life of Benjamin Franklin* (1974) devotes one of its five "scenes" (three out of a total of thirteen chapters) to "Franklin and Electricity." More recently, Ronald W. Clark's *Benjamin Franklin* (1983) opens with a four-page account of the lightning experiments performed in France in 1752 according to Franklin's specifications, which proved that the lightning discharge is an electrical phenomenon and set the seal to his international reputation.[13] Clark has the wit to understand that Franklin's fame would later be a significant factor in the success of his mission to France to procure arms and other equipment for the revolutionary army. Yet it must be confessed that Clark has devoted a mere thirty of five hundred pages to Franklin and the science of electricity, only a little more than Van Doren's twenty-seven pages forty-five years earlier.

The most recent biography, Esmond Wright's *Franklin of Philadelphia* (1986), devotes to science a scant eleven pages out of four hundred.[14] Five of those eleven pages are devoted to nonscientific controversies over the use of lightning rods and peripheral topics, three to lightning experiments, and a bare three pages to theoretical and experimental studies of electricity. Wright gives his readers no idea whatever of any of Franklin's fundamental contributions to electrical science. This example shows that one cannot simply assume that a victory in scholarship is necessarily permanent, that in the case of Franklin historians will universally treat his science fully and with respect.

Among the now recognized features of Franklin's scientific career, a few may merit special attention. Franklin's place in the history of science must be judged by the magnitude of his contribution, not by the number of years he spent on the research. Many historians of the 1930s and earlier would belittle or even dismiss Franklin's scientific work as of no great consequence, primarily because it did not occupy his full-time attention and interest for many years. Franklin was well educated in science, although he had no formal or school training, before he began his studies of electricity. In particular, he was well read in the literature of experimental science and was fully aware of Newton's concept of the aether as a means of explaining attraction; Franklin's own concept of an electrical "fluid" was based on the Newtonian model. Although a self-trained amateur, Franklin could have held his own in science with possessors of advanced degrees.[15] He was the first scientist to win an international reputation in the new branch of science: electricity. His major contribution was the elaboration of a workable theory of electrical "action," one that was remarkably successful in predicting the outcome of experiments of many different sorts. In the opinion of some of his contemporaries, it was this theory that made a "science" of electricity, a subject that had been a congeries of so many "*bizarreries,* or unaccountable phaenomena . . . that a man can scarce assert any thing in consequence of any experiment which is not contradicted by some unexpected occurrence in another."[16]

Franklin's discoveries include the distinction between conductors and insulators, the role of grounding in electrical experiments, the analysis of the Leyden jar, the design of the parallel-plate capacitor, the action of pointed or sharp-edged conductors in "throwing off" and "drawing off" the "electrical fire" or "fluid,"[17] and—the crown of all—the law of conservation of charge. It is to him that we owe some of the commonest words still used in electrical science: plus or positive, minus or negative, electric battery. The reason his terms became the established vocabulary of this science is that his book was so widely read—five editions in English, three in French (in two different translations), and one each in German and Italian—an impressive publication for any scientific book at any time in the modern period.[18]

The law of conservation of charge was of notable significance because it states that in all cases of electrification equal quantities of negative and of positive charge are either produced or disappear. This law implies the transfer or redistribution of electric "fluid" and not the creation of it by rubbing or other modes of charge production. One of the surprising results of Franklin's analysis, through experiments and theory, was that a charged Leyden jar has no more "electricity" in it when charged than when discharged. When the water inside the jar gains a plus (or a minus) charge, the outer coating gains an equal minus (or a plus) charge. Thus, in Franklin's terms, one of these two conductors gains exactly as much electric fluid as the other loses; hence the net amount of electrical fluid is the same in a charged and an uncharged Leyden jar or any other capacitor.

The lightning experiments used to be much misunderstood—and in some respects they still are (see Chapter 6). For it has often not been fully recognized that these experiments were made possible by Franklin's discoveries in pure or basic science: the action of pointed conductors, the function of conductors and insulators, the process of what we call electrostatic induction. The first experiment he designed was the sentry-box experiment. He had not yet performed this experiment himself when he heard that it had been done in France according to his instructions, with a successful outcome. Even before he learned of the results of this experiment, however, he had designed and carried through a somewhat different experiment directed to the same end: the lightning kite. It is often supposed that this experiment must have been very dangerous and that the most remarkable aspect of it is that Franklin was not killed. But the experiment was repeated again and again during the eighteenth century without mishap,[19] although it is true that a German experimenter, Georg Wilhelm Richmann, was killed in St. Petersburg while performing a variation of the sentry-box experiment.

Although Richmann's death shocked many scientists at the time, it later became the occasion for one of Joseph Priestley's better known tongue-in-cheek remarks. In his *History and Present State of . . . Electricity,* Priestley summarized the letter written by Pieter Van Musschenbroek to the naturalist Réaumur, in which the Dutch physicist recounted how he had discovered the principle of the capacitor (which was named the "Leyden jar" after Musschenbroek's hometown).[20] Musschenbroek made the famous statement that he would not be willing to receive such a shock again even if offered the whole kingdom of France. Musschenbroek's letter was published in the *Mémoires de l'Académie Royale des Sciences* and the Leyden experiments became known throughout the scientific world. The discovery had apparently been made when one of Musschenbroek's assistants, Andreas Cunaeus, accidentally formed a capacitor in which he himself was one of the two conductors. This conductor was actually his hand, which held a glass phial of water into which a wire was inserted. The system of wire and water formed the second conductor; the glass of the phial was the insulator or dielectric separating the two conductors. When Cunaeus attempted to remove the wire from the phial after it had been charged, he received a most severe shock. Musschenbroek then exchanged places with him and received an even more powerful jolt, so that—as he said—he thought himself done for, his body being shaken as if by a lightning stroke. Musschenbroek's statement about not ever receiving another such shock from the Leyden jar was the occasion for Priestley's rebuke of his "unworthy sentiment"; Priestley compared Musschenbroek, the "cowardly professor," to the "magnanimous Mr. Boze, who with a truly philosophical heroism, worthy of the renowned Empedocles, said he wished he might die by the electric shock, that the account of his death might furnish an article for the memoirs of the French Academy of Sciences." Then Priestley made his famous comment: "But it is not given to every electrician to die in so glorious a manner as the justly envied Richman."[21]

Franklin's lightning experiments were especially significant in their own day for three reasons that have not been universally appreciated. First, they showed that electrical phenomena occur in nature—and on a large scale. Indeed, it is only in scale that lightning differs from the spark discharges that occur in the experiments made in the parlor or the laboratory. The subject of electricity, in other words, is not merely a kind of "toy" physics for the entertainment of the curious, but a key to some of the most powerful operations of nature.[22] It was a conclusion from Franklin's experiments with lightning and the electrification of clouds that no explanation of the physical world could be considered complete if it did not embrace electricity along with gravity, heat, light, magnetism, and motion.[23] Soon afterward it was found that there are yet other natural sources of electrification. One is the electrification of certain types of crystals, such as tourmaline, when they undergo a change in temperature. A second is the electricity produced by fishes, among them the electric eel and electric torpedo. In these examples of electricity produced by heat or by animal action, there was found to be further confirmation of the generality of the Franklinian law of conservation of charge.

Second, almost at once it was generally recognized that Franklin had provided the first significant confirmation of the Baconian prediction that the pursuit of pure or basic science would lead to major practical innovations. Although men of science had repeated Bacon's statement that knowledge of nature would lead to power and control of nature's forces, there had never been a major example of this process until Franklin's invention of the lightning rod. Nor was there another such outstanding exemplification of the Baconian doctrine—at least one of real practical use—until well into the nineteenth century. As late as 1837, during a French debate on the practicality of government support for pure or basic scientific research, the major example cited was still the Franklinian *paratonnerre.*[24]

Third, and not least in importance, was the significance of the lightning experiments in the warfare against superstition. In those days it was commonly believed that lightning is a manifestation of the wrath of an angry God, or possibly a warning to sinful man, a portent. Church bells were rung during lightning storms in a vain attempt to drive away the destructive force (see Chapter 8). Franklin showed that lightning is in no sense more portentous than any other natural phenomenon. He not only reduced thunder and lightning to natural causes as part of the ordinary operations of nature, but he showed people how to protect themselves (and their homes and barns and churches) from the destructive effects of the lightning discharge.

Franklin was so confident about the outcome of the lightning experiments that he published a description of the lightning rod even before he had experimental proof of the electrical nature of the lightning discharge (see Chapter 6). At that time he thought that the grounded pointed rod would "draw off" the "electrical fire" from passing clouds and silently disarm them so as to prevent a strike, just as in the laboratory a grounded pointed conductor would draw off the charge from a nearby charged body (even though there was no physical contact between the two). In the event, he found that the rods could also perform another function: they would safely conduct the lightning discharge into the ground without damage to a building, its contents, or its inhabitants. Franklin later discovered that clouds are usually electrified negatively rather than positively, so that—as he said in a letter to Collinson in 1753—"for the most part, in thunder-strokes, *'tis the earth that strikes into the clouds, and not the clouds that strike into the earth.*"

The lightning experiments and the invention of the lightning rod gave Franklin at once an international reputation of the highest order. He was awarded the Royal Society's Copley Medal in 1753, was elected a Fellow of the Royal Society in 1756, and in 1773 became a foreign associate of the French Académie des Sciences. Franklin's achievement of a highly successful career wholly in the field of electricity marked the coming-of-age of electrical science and the full acceptance by the scientific community of the new field of specialization.

In the late 1740s, Franklin decided to retire from business and "publick Affairs" in order, as he wrote to Cadwallader Colden, to have "Leisure to read, study, make Experiments, and converse at large with such ingenious & worthy men as are pleas'd to honour me with their Friendship or Acquaintance, on such Points as may produce something for the common Benefit of Mankind, uninterrupted by the little Cares & Fatigues of Business."[25] Toward this end he had put his "Printing house under the Care of my Partner David Hall," had "absolutely left off Bookselling," and had moved to a "more quiet Part of the Town, where I . . . hope soon to be quite a Master of my own Time." Franklin seems to have believed it possible for him to slip back unnoticed into private life.

But, as Carl Van Doren pointed out, "his skill and zeal in recent affairs had made him too well known to be long an undisturbed philosopher." Now, "he was, though not an official, a public man."[26]

From that time until almost the last days of his long life of public service, he continued to yearn for a life of quiet and leisure for the pursuit of science. At the end of the Revolution, Joseph Banks wrote to Franklin to urge him to "abdicate the station of Legislator" and "to return to your more interesting, more elevated, and I will say more useful pursuit of Philosophy." He reminded Franklin that the "head of the Philosopher guides the hand of the farmer to a more abundant crop than nature or instinct or unguided reason could have produced." The philosopher, he went on, "leads the sailor" by inventing tools of navigation and produces other practical innovations.[27] Franklin was no doubt sincere when he replied that he would be "much more happy" to "sit down in sweet Society with my English philosophic Friends, communicating to each other new Discoveries, and proposing Improvements of old ones" than to be in the "Company . . . of all the Grandees of the Earth."[28] On the eve of his return from France to America, he wrote to his friend Jan Ingenhousz that he was at last "a Freeman" after "Fifty Years Service in Public Affairs" and he hoped that Ingenhousz would make him "happy" in "the Little Remainder left me of my Life, by spending the Time with me in America." There, "I have Instruments," he said, "if the Enemy did not destroy them all, and we will make Plenty of Experiments together."[29]

Franklin's correspondence shows that he considered himself to some degree a scientist or natural philosopher on a kind of leave of absence from what he considered to be his fundamental interest. He was active, when opportunity permitted, in some scientific inquiries and he kept up with the progress of science in his own fields of interest. He also took on public or official scientific duties. In England, for example, he served on the Council of the Royal Society and in France he was a leading member of the commission appointed to investigate the claims of Franz Anton Mesmer.

I have mentioned that it used to be thought that Franklin's work in science was of interest not so much for any intrinsic worth in the development of scientific thought as for showing yet another facet of the life and career of one of the founding fathers of our country. It now appears that the very opposite is the case. Franklin's outstanding success in public life in England and in France was considerably aided by his scientific career. Of course, his astonishing achievement at the French court during the war years was due to his skill in diplomacy, his keen sense of timing in public affairs, his shrewdness in matters of human nature and politics, and his skill in playing a kind of poker for world stakes. But there can be no question that when Franklin arrived in France on that grave mission he was already a public figure, well known to the French court and to the French public at large in a sense that would have been true of no other American in the political diplomatic arena, and this was so because of his stature as a scientist and because of the spectacular nature of his work on lightning. Of whom else could it have been said by Turgot, "Eripuit coelo fulmen" (He snatched the lightning from the sky)?[30]

It is true that Franklin's experiments and his theory of electricity achieved major status after the spectacular lightning experiments and the successful demonstration of the utility of the lightning rod. But the real test of Franklin's ideas, and the only true gauge of the worth of his theory, was whether or not those who went along with the concept of a single fluid theory of electricity could explain and predict the outcome of laboratory experiments and the great phenomena of nature. In 1747, just before he came across Franklin's work, Buffon referred to the low state of electrical knowledge—not really as yet established as a science—in the following terms, as reported by John Turberville Needham: "he thought the whole subject of electricity, though illustrated with so great a variety of experiments, very far from being yet sufficiently ripe for the establishment of a course of laws, or indeed of any certain one, fixed and determined in all its circumstances."[31] How pleased he was with Franklin's theory can be seen in the fact that he sponsored a translation of Franklin's *Experiments and Observations on Electricity* into French.[32] He was also delighted, we may be sure, that he was at the same time delivering a blow to the Abbé Nollet, then the reigning authority on electricity in France; Nollet was the disciple and protégé of Réaumur, with whom Buffon was then having a fierce and acrid controversy.[33] Twenty years later, Jacques Barbeu-Dubourg could still stress the value of Franklin's theory (especially as compared with Nollet's) in successfully coordinating and predicting the outcome of experiments. Barbeu-Dubourg wrote that Franklin says, "Do this, and that is what will happen; change this circumstance, and that is what will be the result."[34] In Italy, Father Beccaria showed how "the truth of the Franklinian theory" is established by appealing to "the two masters of true learning, observation and experiment."[35]

Almost two hundred years later, J. J. Thomson, discoverer of the electron, repeated the sentiments of Barbeu-Dubourg almost word for word as he explained that we still account for the major facts of electrostatics in very nearly the way Franklin proposed. After describing the general character of Franklin's one-fluid theory, Thomson said: "The service which the one-fluid theory has rendered to the science of electricity, by suggesting and co-ordinating researches, can hardly be overestimated. It is still used by many of us when working in the laboratory. If we move a piece of brass and want to know whether that will increase or decrease the effect we are observing, we do not fly to the higher mathematics, but use the simple conception of the electric fluid which would tell us as much as we wanted to know in a few seconds."[36] Thomson's version of the one-fluid theory (the one still in use) differs from Franklin's in one major respect, however, for it includes an additional postulate introduced by Aepinus (see note 39).

On the purely theoretical side, Franklin's really significant achievements were his explanations of the variety of charge production by what is known as electrostatic induction. Closely allied to these explanations was his elucidation of the phenomena of the Leyden jar. Franklin's explanation required that "electrical fluid" not permeate the glass of which the jar was composed, although its influence could pass through the glass. Thus if a positive charge (additional electrical fluid) were given to the inner conductor (that is, to the water inside the jar and the wire going down through the cork into the wire), then by repulsion some of the normal electrical fluid on the outer conductor (the metal coating of the jar) would be driven away—if that outer conductor were grounded—so as to produce a negative charge on the outer conductor simultaneously with the positive charge to the inner conductor. Unless the outer conductor were grounded, even momentarily, or a spark were drawn, the inner conductor could not gain its great positive charge. By symmetry, it was obvious to Franklin and his followers, the positive charge could be given by contact to the outer conductor, and in this case the grounded inner conductor would gain a negative charge by electrostatic induction. In Franklin's explanation of the Leyden jar, the principles of electrostatic induction are set forth plainly; they were elaborated for other situations in a paper by John Canton and by Franklin himself in his addenda to Canton's explanations.[37] Franklin's analysis of the Leyden jar did not stop with his account of the simultaneous production of opposite charges on the inner and outer conductors. He showed that the charge "resides" in the glass, not on the conductors as individual entities, a phenomenon we know today under the somewhat forbidding name of polarization of the dielectric. He also demonstrated that the charges on the two conductors are always equal, though of opposite magnitude, an exemplification of what we know today as the Franklinian principle of the conservation of charge.

Despite the great success of Franklin's one-fluid theory, it did not gain full endorsement by everyone. A major deficiency of the theory was its failure to explain why two negatively charged bodies should repel each other, as experiment shows they do.[38] This deficiency was rectified by Aepinus, who modified the Franklinian theory by adding the supposition that matter deprived of its electric fluid will repel other such matter. Aepinus, however, did not invent this modification primarily in order to account for the mutual repulsion of negatively charged bodies, but rather to remedy another—and more fundamental—weakness in the theory. For when he added up the forces of attraction and repulsion ("common" matter attracting common matter, electric matter repelling electric matter, and common matter attracting and being attracted by electric matter), he found that simple mathematical considerations required an additional set of mutual repulsions.[39]

The failure of Franklin's theory, uncovered by Aepinus, was mathematical and somewhat subtle and difficult to grasp. But, as we shall see, there was another basic problem that was easier to comprehend. In order to gain a full picture of Franklin's scientific stature and influence we must take account of such failures of his theory as well as the successes. In this chapter, as elsewhere in the book, I have stressed Franklin's positive contributions to electrical science—both as they were appreciated by his contemporaries and as they appear in retrospect to critical historians of science—since a primary purpose has been to educate American historians (and even biographers of Franklin) who either were ignorant of Franklin's place in the history of science or shared misconceptions about science, its history, and Franklin's achievement. Even historians of science have often shown an ignorance of Franklin's contributions to electricity. I have, accordingly, not explored here the weaknesses of Franklin's theory that led to successive modifications by Franklin himself and others, and its eventual abandonment in favor of a two-fluid theory.[40]

I hope that no reader will conclude that Franklin's theory had no flaws or that there existed in Franklin's day no rival to challenge his intellectual creation. Despite some assertions to the contrary, Franklin's theory has not survived unchallenged and unaltered to our day. This is hardly the rule in the history of scientific theories, especially in those branches of science in which new knowledge is being rapidly accumulated by experiment.

Theories in science may be considered successful not merely for their ability to coordinate and predict experimental results and to suggest new experiments, but also for their role in providing a platform on which new and even more successful theories may be built. A measure of a theory's success may, therefore, be its importance in advancing a science to the next theoretical stage. Franklin's theory of electrical action was superseded by a two-fluid theory, but it is to be noted that each of these fluids was to a considerable degree a Franklinian fluid, acting in accordance with the most basic Franklinian postulates and principles (such as the conservation of charge). Even the names Franklin gave to the two states of electrification—plus and minus—have outlasted the original physical significance he gave to them.

As has been mentioned, one of the strengths of Franklin's theory was its success in giving scientists a basis for successfully predicting the outcome of many experiments and especially in providing an explanation for the action of the Leyden jar, the most challenging electrical phenomenon of the day. His theory and concepts attained a special importance after the lightning experiments and the invention of the lightning rod. But the chief rival theory (proposed by the Abbé Nollet) was also to some degree successful, especially in regard to certain phenomena for which Franklin and his followers were not able to provide easy explanations. Its chief failure was that it could not account for the phenomena of the Leyden jar in the way that Franklin's theory could.[41]

The chief and obvious weakness of Franklin's theory lay in the doctrine of "electrical atmospheres," a supposed cloud of electrical fluid that was thought to envelop bodies having a positive charge. This part of the theory encountered real difficulties, requiring the introduction of ad hoc hypotheses (by Franklin and others), notably in order to explain certain experiments of John Canton.[42] The difficulties that arose in connection with Franklin's doctrine of electric atmospheres have been studied by Roderick W. Home, who has also explored the basic mathematical flaw in Franklin's postulates that was corrected by Aepinus.[43] The virtues of the rival theories to Franklin's, especially the ideas of Nollet, have been explored by John Heilbron, who has also delineated the stages whereby Franklin's theory was radically altered by such of his followers as Canton and Volta and was eventually replaced.

Many nineteenth-century theorists were not very happy with the two-fluid theory of electricity, in which there was supposed to be one fluid for "resinous" (or negative) electricity and another for "vitreous" (or positive) electricity.[44] By the time of J. Clerk Maxwell, later on in the nineteenth century, it came to be believed that the concept of fluid in any physical sense was misleading. But at the turn of the century, when J. J. Thomson discovered the electron, it seemed as if Franklin had been on the right track after all (at least with the modification by Aepinus). R. A. Millikan, who determined the charge on the electron and who provided additional evidence to support the corpuscularity of electric charges, even went so far as to hail Franklin as the discoverer of the electron.[45] Historians of science at present use the term "Whig history" to characterize such an evaluation of Franklin's science in terms of the accidental apparent coincidence of his concepts and those held by physicists at some point in the twentieth century.[46] They would rather see Franklin's importance in its historical dimensions, in the degree to which his experiments and theory moved the subject of electricity forward, introducing a major intellectual leap in the progress of knowledge.

2

Franklin's Scientific Style

Benjamin Franklin's preparation for a career of scientific research included a careful and profound study of the ideas and principles of investigation of Isaac Newton. It is, however, one of the paradoxes of Benjamin Franklin's career that his contemporaries considered him to be a foremost Newtonian scientist even though he did not have the skill or training to be able to read Newton's *Principia*. His Newtonian science was derived from the *Opticks*, which for eighteenth-century experimental scientists was a manual of the experimental art.[1]

The Principia *and the* Opticks

In recent decades we have become aware that Newton's two great works—the *Principia* and the *Opticks*—became the foundations of very different and even distinct scientific traditions, derived from their subject matter, formal style, and mode of investigation and proof.[2] The *Principia* was devoted to rational mechanics (a name introduced by Newton: "mechanica rationalis"[3]) and its elaboration in celestial dynamics and in the Newtonian "system of the world." This domain of science was traditionally developed mathematically, proposition by proposition, from an initial set of definitions and axioms, which in the *Principia* were the Newtonian laws of motion. It was befitting this rigorous and austere mathematical style that the *Principia* was written in traditional Latin.

The *Opticks* was a quite different sort of book and accordingly engendered a somewhat separate tradition in science. Written in a graceful English rather than in the severe Latin of the *Principia,* the *Opticks* was almost totally devoid of mathematics.[4] That is, the *Opticks* did not develop its subject in a systematic manner by mathematical proofs (in the style of the *Principia*) but rather stated its propositions and then introduced "The Proof by Experiments." In actual fact, not once in the *Opticks* does Newton develop proofs or discussions in a complete sequence, using the recognized tools of mathematics, that is, equations, proportions, trigonometric identities, fluxions (or the calculus), and the like. Indeed, in book 2 Newton notes in passing that "all these things follow from the properties of Light by a mathematical way of reasoning," but "the truth of them may be manifested by Experiments."[5]

In attempting to define the character of the *Opticks,* however, we must be careful not to confuse "mathematical" in this sense with "numerical" or "quantitative" or "exact." The *Opticks* is a quantitative work in that it is based on numerical results of measurements—for example, of the width of "Newton's rings" or various examples of refraction. Furthermore, the experiments lead to quantitative laws, such as the law of refraction, which are expressed in mathematical language (the sine of angles of incidence and of refraction). But such laws do not appear as consequences of a mathematical structure, as derivations from mathematical axioms, nor are they proved by the use of mathematical techniques. Nor, for that matter, does Newton in the *Opticks* go on to develop mathematically further propositions based upon such laws.

The result is that the *Opticks* was a work readily accessible to experimental scientists such as Benjamin Franklin, Stephen Hales, Joseph Black, and A.-L. Lavoisier, who did not need to be trained in higher mathematics to understand its contents. That is, the *Opticks* did not demand of its readers a knowledge of the theory of proportions, the geometry of conic sections, the theory of limits, fluxions or the infinitesimal calculus, or skill in using infinite series, such as would be needed to follow the argument of the *Principia.*[6]

I may note in passing that it was not Newton's original intent to develop this subject in a manner or style so different from that of the *Principia.* We have evidence aplenty, analyzed by Alan Shapiro and Zev Bechler, that Newton tried again and again to find a basis for producing a mathematical science of optics, but all of his efforts ended in failure.[7] After years of hesitation and self-doubt, Newton agreed in 1704 to present to the world of science the results of his extensive and original experimental investigations in the domain of light and color. Thus he could guarantee his claims to priority in discovery. But the form of the *Opticks*—in English rather than in Latin and without mathematics—was, I believe, in a sense a confession of his failure. This is no doubt the reason why the title page of the *Opticks* does not bear the author's name.[8] That this omission was the result of design and not of accident can be seen in the fact that in at least one copy of record, the name of the author has been inserted on a specially printed title page pasted onto the stub of the originally canceled or cut-off page.[9]

The "Queries" of the Opticks

The popular appeal of the *Opticks* was enhanced by the famous "Queries" which formed its conclusion. These dealt with a large variety of experimental and theoretical topics, including physical optics, vision, fire or combustion, radiant heat, electricity, magnetism, corpuscular or atomic physics, the phenomena and theory of waves, chemistry, the nature of matter, and even a possible cause of universal gravity. The first edition (1704) contained only sixteen Queries, which were rather short, but in later editions (Latin, 1706; revised English edition, 1717/1718) their number was successively increased to an eventual thirty-one, the later ones occupying many pages each.[10] These Queries were not questions in the ordinary sense. All but one show their rhetorical nature by being in the negative. That is, Newton does not ask in a truly querulous fashion whether flame may be "a vapour, fume or exhalation heated red hot"; rather, he insists (Query 10): "Is not Flame a Vapour, Fume or Exhalation heated red hot, that is, so hot as to shine?" That Newton knew the answer is evident at once in the ensuing discussion in which he presents the evidence to support his position. The form of query, in other words, was a rhetorical device so that Newton could freely set forth his ideas even on topics for which he had no mathematical proof or for which the experimental evidence was not overwhelmingly definitive. He introduced this section of the *Opticks* with the frank declaration that there were important subjects which he had hoped to pursue, but could no longer investigate. He would, accordingly, conclude his treatise "with proposing only some Queries in order to [provoke] a farther search to be made by others."[11]

Newtonians had no doubts concerning Newton's own beliefs in relation to each of the topics developed in the Queries. Stephen Hales, for one, said simply that in the latter part of the *Opticks* Newton had explained many things "by way of query." J. T. Desaguliers, for another, particularly lauded the Newton of the Queries, which he said contain "an excellent body of philosophy."[12] The form of question, he explained, had nothing to do with uncertainty about truth but arose merely from Newton's "modesty," since Newton would never assert as true any proposition that he "could not prove" by "mathematical demonstrations or experiments."

In this sentiment Desaguliers was merely restating what Newton himself had said in the "Advertisement" to the second edition of the *Opticks*. In order "to shew that I do not take Gravity for an essential Property of Bodies," Newton wrote, "I have added one Question concerning its Cause." He had chosen "to propose it by way of a Question," he continued, "because I am not yet satisfied about it for want of Experiments."

In the Queries Newton had made an inventory of problems requiring experimental investigation and of areas to be explored by future investigators. Hence the Queries displayed a research program for those scientists of the eighteenth century who conceived that a major mode of advance in science lay in the use of experiments rather than mathematical derivations and mathematical solutions to problems. These scientists worked in such areas as optics, plant and animal physiology, heat, electricity, magnetism, and chemistry and the structure of matter.

Newtonian Experimental Natural Philosophy

My own recognition of the significance of the *Opticks* in relation to the rise of the experimental science of the eighteenth century came about years ago during my reading of Benjamin Franklin's *Experiments and Observations in Electricity*[13] and my attempts to understand Joseph Priestley's comparison of Franklin's "true principles of electricity" with Newton's "philosophy" and "the true system of nature in general."[14] Eventually, I became aware of Hélène Metzger's writings, proposing the existence of a Newtonian tradition stemming from the *Opticks* rather than the *Principia*.[15] Her works documented this influence of the *Opticks* in a chemical context; later I came upon a briefer suggestion of this thesis (again in a chemical context) by Pierre Duhem.[16] At this time (1940–41), while I was preparing for the press my edition of Franklin's book on electricity,[17] the nature and influence of Newton's *Opticks* became a subject of almost constant discussion between me and my fellow graduate student Henry Guerlac, who was then already deep in his studies of Lavoisier and his antecedents, a joyful complement to my own explorations of physics in that same era.[18] As a result it would be well-nigh impossible for either of us to be absolutely certain how much came from the other, or how much we both derived from Hélène Metzger or from our common mentor Giorgio de Santillana.

A number of different kinds of evidence gave support to a difference in appeal of the *Principia* and the *Opticks* and, consequently, a different scientific tradition allied with these two works. For example, I discovered that John Locke, a thoroughgoing Newtonian in philosophy, found the *Principia* wholly beyond his comprehension and appealed to Christiaan Huygens to ascertain whether "all the mathematical *Propositions* in Sir

Isaac's *Principia* were true." On "being told he might depend on their Certainty," J. T. Desaguliers reported, Locke "took them for granted and carefully examined the Reasonings and *Corollaries* drawn from them." Desaguliers contrasted Locke's approach to the *Principia* with the way in which "he read the *Opticks* with pleasure, acquainting himself with every thing in them that was not merely mathematical."[19] The story of Locke's attempts to read the *Principia* had been told to Desaguliers "several times by Sir *Isaac Newton* himself."[20] This general appeal of the *Opticks* can also be seen in the fact that Francesco Algarotti's popular book on Newtonianism (oft reprinted both in the original Italian and in translation[21]) was devoted entirely to the *Opticks*. Pieter van Musschenbroek, one of the great Dutch experimental scientists of the eighteenth century, and a discoverer of the principle of the capacitor (the Leyden jar), held that "one should have continually before one's eyes those two perfect models that the two great men of the century have left us, that is to say, the *Opticks* of Newton and the *Chemistry* of Boerhaave."[22]

An example of the influence of the *Opticks* is afforded by Stephen Hales's *Vegetable Staticks* (1727), a work still ranked among the classics of experimental science. [23] Hales was a pioneer in the application of quantitative methods to biology. In *Vegetable Staticks* he explored such topics of quantitative biological experiment as leaf growth, root pressure, and plant nutrition, and the quantitative study of gases. In a companion volume, *Haemastaticks,* Hales described the first measurements ever made of blood pressure in animals.[24] The text of *Vegetable Staticks* contains seventeen mentions of Newton, of which fifteen are either quotations from or references to the *Opticks*. The other two are concerned, respectively, with Newton's method of calibrating thermometers and with his essay on acids.[25]

As I began to accumulate evidence of the existence of the two rather separate Newtonian traditions, it occurred to me to see what the great dictionaries and encyclopedias in the age of Newton had to say about the varieties of Newtonianism and the Newtonian philosophy. To my astonishment and delight I found that Harris, Chambers, and the great *Encyclopédie* all indicated that there were several different senses then current.[26] The primary ones were those that I had associated with the *Principia* and with the *Opticks*. But it was only much later that I learned that the fundamental differences between these two works had been a matter not of choice so much as of necessity.[27] True, the *Opticks* would have in any case depended more heavily on actual experiments; but the essential qualities of speculation, physical insight, induction, and non-mathematical inference which characterize the *Opticks* were the result less of original design or intent than of the stark reality of mathematical failure.

The principles of the Newtonian experimental philosophy were set forth in a series of first-rate nontechnical expositions—Henry Pemberton's *View of Sir Isaac Newton's Philosophy,* Herman Boerhaave's *New Method of Chemistry,* W. J. 'sGravesande's *Natural Philosophy,* J. T. Desaguliers's *Experimental Philosophy,* Hales's *Statical Essays,* and others. These works were read or studied by Franklin in the late 1730s and early 1740s, the years of self-education in the sciences and preparation for a career as a scientist.[28] These books brought Franklin into contact with the experimental art and, in particular, with Newtonian atomism and the concept of a subtle and imponderable medium or fluid, manifested in heat, optical phenomena, and electrical effects—an aether or aetherial medium composed of particles that repel one another. This was a model for Franklin's concept of an electrical fluid. In 'sGravesande's book, for example, Franklin would have found a discussion of fire particles attracted by bodies, uniting to the corpuscles composing bodies, that—while contained in all bodies—do not enter into all bodies with equal facility.

Desaguliers wrote about magnetism in a way that could apply equally to electricity. If "ever we come to know the causes of the various operations of magnetism," he declared, it "will be sooner owing to a comparison of the experiments and observations" of good inquirers than "to twenty hypotheses of men whose warm imaginations supply them with what may support their solutions, while daily observations and common laws of motion can easily confute them."[29] Like other Newtonians, Desaguliers introduced electricity in order to show that "attraction" actually occurs in nature, however "occult" may be its cause. But electrical phenomena also provided an example to illustrate a challenging position set forth by Newton in the Queries of the *Opticks:* the possibility of "attraction and repulsion in the same body at considerable distance," made "evident in several electrical experiments."[30] The last five pages of Desaguliers's book on experimental Newtonian science (1734) were devoted to an exposition of electricity which may have been Franklin's first introduction to that subject.[31] Here Desaguliers adapted and expanded Newton's fundamental concept of mutual particulate repulsion, that (Query 21) "aether (like our air) may contain particles which endeavour to recede from one another." This concept was basic to Desaguliers's thoughts on elasticity, Hales's on gases, Franklin's on the electric fluid, Lavoisier's on caloric, and—later—Dalton's on atoms.

Not only did Desaguliers give a succinct summary of electrical experiments made in the 1720s and 1730s;[32] he also included in the second volume an English version of his *Dissertation concerning Electricity* (1742), which had won the prize of the Academy of Bordeaux. He had been a younger member of Newton's circle and he was the author of a standard manual of Freemasonry, which Franklin published in an American edition.[33] Desaguliers was, therefore, an especially important author for Franklin in the 1740s. Furthermore, Franklin knew Desaguliers's writings on stoves and chimneys and his English version of Gauger's book on these topics.[34] Accordingly it is significant that Desaguliers not only set forth principles of elastic imponderable fluids that formed a conceptual matrix for Franklin's idea of the electric fluid; he also—by including an essay on electricity in a treatise on Newtonian physics—declared that the nascent science of electricity was a legitimate topic of inquiry for the Newtonian experimental philosopher.

Desaguliers made a number of references to the work of Hales, notably his research on the chemistry and physics of gases and his conviction that the Queries contained scientific truths and sound guides to scientific practice. Hales was a notable author for Franklin in the decades before his electrical research, and his correspondence shows how deeply he had read in Hales's books. Franklin would have been impressed by the way Hales drew on the authority of the Queries of Newton's *Opticks* for the existence of "active aerial particles repelling each other." Hales's presentation, like those of other authors Franklin was reading, gave Franklin a concept of elastic fluids which might exist in and around solid matter, which became basic to the Franklinian theory of electricity. Hales, furthermore, introduced the subjects of electricity and lightning into both of his major works, *Vegetable Staticks* and *Haemastaticks*. Franklin would have found a great maxim for experimental science in Hales's expression of hope that he "could be as happy in drawing the proper inferences" from experiments as in performing and reporting them.

From such works Franklin learned the Newtonian principles of particulate attraction and repulsion; he also became familiar with the concept of an elastic fluid.[35] Above all he learned the Newtonian respect—stated explicitly in precepts by 'sGravesande, Desaguliers, and Hales—for carefully performed experiments that must be accurately reported.

The Franklinian Theory of Electricity

Franklin's friend, the naturalist Buffon, once wrote that style is the man himself. Franklin is celebrated for a style of life and action and a style of prose. It is more for its style than its content that his autobiography is generally reckoned to be an American classic. In his report "Opinions and Conjectures," and in his other writings on electricity, and notably in his account of the experiment of the dissectible condenser (or capacitor), we may see Franklin as a master of expository prose. His skill in writing effective, straightforward prose is also manifested in his account of the Pennsylvania Hospital (see Chapter 10). As we read Franklin's presentation of his theory and experiments, we may nod in agreement with Humphry Davy, who told students of agricultural chemistry to take Franklin's book on electricity as a model, since "the style and manner of his publication on electricity are almost as worthy of admiration as the doctrine it contains." Science, Davy concludes, appears in Franklin's language "in a dress wonderfully decorous, the best adapted to display her native loveliness."[36]

Franklin's style as a writer was matched by a style and skill in making experiments and drawing conclusions from them. The record of Franklin's initial electrical discoveries is contained in a series of epistolary reports sent by him to England on 25 May 1747, 28 July 1747, and 29 April 1749 (two letters), leading up to a formal presentation of the "Properties and Effects of the electrical Matter, arising from Experiments and Observations, made at Philadelphia, 1749."[37] The heading of this paper was the source of the title given to his book.

In this 1749 summary Franklin shows the influence of ideas put forth by Newton in Queries 18–24 of the *Opticks,* concerning "a much subtiler Medium than Air," an "aethereal Medium." This aether, according to Newton, pervades and resides in the pores of bodies and varies in density from one body to another. We have seen that it is very "rare" (Query 21) and may ("like our Air") be composed of "Particles which endeavour to recede from one another." Because of this property or force of mutual repulsion, this "aether" constitutes an "elastick" fluid. The particles are characterized by an "exceeding smallness," so that it is by many orders of magnitude "more rare" than "our Air" and so can easily penetrate into ordinary bodies.

With some important alterations or modifications this is Franklin's electric fluid. From his postulates it is apparent that, like the aetherial medium whose properties Newton developed in the Queries of the *Opticks,* Franklin's electric fluid is subtle and particulate, composed of particles that repel one another. Postulate 1 reads: "The electrical matter consists of particles extremely subtile, since it can permeate common matter, even the densest metals, with such ease and freedom as not to receive any perceptible resistance."[38] This primary statement is at once qualified by postulate 3, about the mutual repulsion of the particles: "Electrical matter differs from common matter in this, that the parts of the latter mutually attract, those of the former mutually repel each other."

At this point Franklin added a new and original additional property, expressed in postulate 4: that the particles of electrical matter "are strongly attracted by all other matter." These postulates accord with the general principle that all electrical phenomena are the result of a change in the distribution of the electrical matter in bodies or of some net change in the total quantity of electrical matter in individual bodies.

This principle is embodied in a more general principle which we know as "conservation of charge": that "electricity" is never created or destroyed in electrical phenomena, that electrical phenomena are caused by changes in either the distribution or the net quantity of electrical matter in bodies. In experiment after experiment, Franklin showed that the production of a positive charge in one body (a net gain in electrical fluid) is always accompanied by an equal and opposite negative charge (a net loss in electrical fluid) in one or more other bodies. Other experiments showed in a remarkable way how electrical effects may be produced by the temporary change in the normal condition (or equal distribution) of the electrical fluid in one or more bodies. We know this as electrostatic induction.

The doctrine of conservation had been a hallmark of Newtonian rational mechanics. The first "axiom" or "law of motion" in the *Principia* declared the property of inertia—the conservation of a body's state of motion or of rest. A basic principle enunciated in the scholium following the laws of motion was the conservation of momentum.

In his first report on his research, Franklin explained how he and his fellow experimenters, Philip Syng and Thomas Hopkinson, had come to the conclusion "that the electrical fire was not created by friction, but collected, being really an element diffused among, and attracted by other matter, particularly by water and metals."[39] Franklin described the fundamental experiments leading to that conclusion as follows:

> 1. A person standing on wax, and rubbing the tube, and another person on wax drawing the fire, they will both of them, (provided they do not stand so as to touch one another) appear to be electrised, to a person standing on the floor; that is, he will perceive a spark on approaching each of them with his knuckle.
>
> 2. But if the persons on wax touch one another during the exciting of the tube, neither of them will appear to be electrised.
>
> 3. If they touch one another after exciting the tube, and drawing the fire as aforesaid, there will be a stronger spark between them than was between either of them and the person on the floor.
>
> 4. After such strong spark, neither of them discover any electricity.

The Franklinian explanation appears in the straightforward and clear unadorned prose that was the ideal of the Newtonian scientist. (The complete statement is given at the end of Chapter 5.) The general significance of the new electrical theory was—as Franklin wrote in a letter to Cadwallader Colden in 1747—"that the Electrical Fire is a real Element, or Species of Matter, not *created* by the Friction, but *collected* only." Here was the first stage of the Franklinian revolution in electricity.

New theories or new paradigms, as T. S. Kuhn has stressed, require a new language, which eventually leads to a degree of incommensurability between an older paradigm and a new one. Franklin has described the "new terms" that had arisen among the Philadelphia experimenters. These were soon to become part of the standard vocabulary of electrical science, along with some other terms such as "electrical battery." So strongly did the Franklinians influence the course of science that even when the single-fluid theory gave way (temporarily) to a two-fluid theory, the two fluids were known by the Franklinian names of positive and negative. Even today, long after the discovery of the direction in which a current of electrons moves, we still analyze electric circuits in terms of a "fictitious" (Franklinian) current moving in the opposite direction.

Franklin's first report also contained an account of a major pair of discoveries: "the wonderful effect of pointed bodies, both in *drawing off* and *throwing off* the electrical fire."

> Place an iron shot of three or four inches diameter on the mouth of a clean dry glass bottle. By a fine silken thread from the ceiling, right over the mouth of the bottle, suspend a small cork-ball, about the bigness of a marble; the thread of such a length, as that the cork-ball may rest against the side of the shot. Electrify the shot, and the ball will be repelled to the distance of four or five inches, more or less, according to the quantity of Electricity.
>
> When in this state, if you present to the shot the point of a long, slender, sharp bodkin, at six or eight inches distance, the repellency is instantly destroyed, and the cork flies to the shot. A blunt body must be brought within an inch, and draw a spark to produce the same effect. To prove that the electrical fire is *drawn off* by the point, if you take the blade of the bodkin out of the wooden handle, and fix it in a stick of sealing-wax, and then present it at the distance aforesaid, or if you bring it very near, no such effect follows; but sliding one finger along the wax till you touch the blade, and the ball flies to the shot immediately.[40]

It is to be noted that in this remarkable experiment Franklin has shown, not only that metallic (or conducting) points have the ability to "draw off" the charge (excess electric fluid) from an insulated, positively electrified conductor, but that this ability requires that the point must be grounded. On a small scale this experimental discovery provided the model for Franklin's later experiment of the sentry box, in which a tall, grounded, pointed conductor would be able to "draw off" the electric charge from clouds—if clouds are electrified and if, therefore, the lightning discharge is an electrical phenomenon.[41] We must note, finally, the clarity and direct simplicity of the Franklinian exposition.

Franklin also reported that "points will *throw off* as well as *draw off* the electrical fire." Lay "a long sharp needle upon the shot, and you cannot electrise the shot so as to make it repel the cork-ball." In another version of the experiment a needle was fixed "to the end of a suspended gun-barrel, or iron-rod, so as to point beyond it like a little bayonet." Under these circumstances, "the gun-barrel, or rod, cannot by applying the tube to the other end be electrised so as to give a spark, the [electrical] fire continually running out silently at the point."[42] This latter experiment had been designed by Thomas Hopkinson, who had supposed that he would be able to draw "a more sharp and powerful spark from the point, as from a kind of focus." He "was surprized to find little or none."[43] Here we see an example of a common phenomenon in science, in which the failure of an experiment provides the key to a wholly unexpected discovery.

Franklin and the Leyden Jar

One of the most spectacular examples of the application of the Franklinian doctrines was in Franklin's analysis of the Leyden jar, the first condenser or capacitor. Basically such a device consisted of a nonconductor or insulator (the glass bottle) placed between two conductors, which could be the metal outer coat of the bottle, coming about two-thirds of the way to the top, and the inner water (or lead shot), connected by a wire or chain to a metal rod and knob set in a cork in the neck of the bottle. The jar was charged by bringing the knob near or in contact with one of the "prime conductors" of an electrostatic generator while the coating was temporarily grounded, for example, by contact with the experimenter's body.

Among the mysteries of the jar in Franklin's day was the fact that it could apparently acquire or hold a much greater charge than would have been expected, much more than an ordinary insulated conductor of the same size. The discharge seemed to have a power or force that was beyond comprehension—enough to cause hundreds of persons holding hands to leap into the air simultaneously.

In ordinary use the jar was charged through the knob while the coating was grounded, as by being held in an experimenter's hand. Franklin later reported that the jar could equally be charged through the outer coating, if the knob were grounded.[44] The force of the jar's electricity was manifested when a circuit was made between the outer coating and the inner system of conductors, which consisted of the knob and rod, the wire or chain, and the water. For instance, if an experimenter were to charge a jar held in one hand and would then try to remove the knob-rod-chain with his other hand, the discharge would take place through his body; the force of the discharge could be so severe as to produce great chest and shoulder pain and even unconsciousness. Such was the experience of Pieter van Musschenbroek, one of the inventors of the Leyden jar.[45]

Franklin's analysis of the Leyden jar shows at once his supreme skill as an experimenter and the way in which he applied the Newtonian principle of the electrical fluid to explain his observations. By a careful set of experiments he showed that the positive charge on the outer coating is equal in magnitude to the negative charge on the inner conductors. A particularly ingenious set of experiments made use of a small ball suspended on a silk string, which could swing back and forth between the knob and a wire fastened to the outer conductor. In this way Franklin was able to produce a discharge in a series of small or gradual steps rather than a single instantaneous explosion.

Since the charge of the inner and the outer conductors of a charged jar proved to be equal and opposite, Franklin was led to the odd conclusion that when the jar was charged, one conductor would have lost just as much electric fluid as the other had gained—so that the net amount of fluid in a charged jar was exactly the same as in an uncharged jar. He wrote: "When we use the terms of *charging* and *discharging* the phial, it is in compliance with custom, and for want of others more suitable," as "we are of opinion that there is really no more electrical fire in the phial after what is called its *charging,* than before, nor less after its *discharging.*"[46]

Franklin made another wholly unsuspected discovery about the Leyden jar, namely, that "the whole force of the bottle, and power of giving a shock, is in the GLASS ITSELF; the non-electrics [i.e., conductors] in contact with the two surfaces, serving only to *give* and *receive* to and from the several parts of the glass; that is, to give on one side, and take away from the other."[47] The record of the experiment yielding this new information shows Franklin's mastery of the experimental art.

> Purposing to analyse the electrified bottle, in order to find wherein its strength lay, we placed it on glass, and drew out the cork and wire which for that purpose had been loosely put in. Then taking the bottle in one hand, and bringing a finger of the other near its mouth, a strong spark came from the water, and the shock was as violent as if the wire had remained in it, which shewed that the force did not lie in the wire. Then to find if it resided

in the water, being crouded into and condensed in it, as confined by the glass, which had been our former opinion, we electrified the bottle again, and placing it on glass, drew out the wire and cork as before; then taking up the bottle, we decanted all its water into an empty bottle, which likewise stood on glass; and taking up that other bottle, we expected, if the force resided in the water, to find a shock from it; but there was none. We judged then that it must either be lost in decanting, or remain in the first bottle. The latter we found to be true; for that bottle on trial gave the shock, though filled up as it stood with fresh unelectrified water from a tea-pot.

To find, then, whether glass had this property merely as glass, or whether the form contributed any thing to it; we took a pane of sash-glass, and laying it on the hand, placed a plate of lead on its upper surface; then electrified that plate, and bringing a finger to it, there was a spark and shock. We then took two plates of lead of equal dimensions, but less than the glass by two inches every way, and electrified the glass between them, by electrifying the uppermost lead; then separated the glass from the lead, in doing which, what little fire might be in the lead was taken out, and the glass being touched in the electrified parts with a finger, afforded only very small pricking sparks, but a great number of them might be taken from different places. Then dextrously placing it again between the leaden plates, and compleating a circle between the two surfaces, a violent shock ensued.

Which demonstrated the power to reside in glass as glass, and that the non-electrics in contact served only, like the armature of a loadstone, to unite the force of the several parts, and bring them at once to any point desired: it being the property of a non-electric, that the whole body instantly receives or gives what electrical fire is given to or taken from any one of its parts.[48]

One feature of Franklin's experiment was to show "the power to reside in glass as glass." We must recall that he and his fellow experimenters had already shown that some electrical properties depend on shape and not just substance. An example was the action of pointed as opposed to blunt conductors. So we may understand why it was important for Franklin to produce a capacitor of a wholly different shape from a bottle—the first parallel-plate condenser or capacitor.[49] He also combined several such capacitors to make what "we called an *electrical-battery*."[50]

A feature of Franklin's analysis of the Leyden jar was that the "equilibrium cannot be restored in the bottle by *inward* communication or contact of the parts." To discharge a jar, or to restore equilibrium, "a communication" must be "formed *without* the bottle between the top and bottom, by some non-electric, touching or approaching both at the same time." This astonishing result led Franklin to wonder how this could occur. "These two states of Electricity," he wrote, "the *plus* and *minus*," are "combined and balanced in this miraculous bottle" in "a manner that I can by no means comprehend." He showed how different this phenomenon was from what one might have expected by analogy with the effects of air pressure:

> If it were possible that a bottle should in one part contain a quantity of air strongly comprest, and in another part a perfect vacuum, we know the equilibrium would be instantly restored *within.* But here we have a bottle containing at the same time a *plenum* of electrical fire, and a *vacuum* of the same fire; and yet the equilibrium cannot be restored between them but by a communication *without!* though the *plenum* presses violently to expand, and the hungry vacuum seems to attract as violently in order to be filled.[51]

In Franklin's theory of electrical action, the accumulation of excess charge to produce positive electrification produced an "atmosphere" of electric fluid extending out into the space around a body. These atmospheres enabled him to explain how bodies might interact "at a distance" and produce repulsion between two positively charged bodies not in contact and similarly produce attraction between a negatively and a positively charged body. This doctrine of electrical atmospheres proved to be a major weakness in the Franklinian theory. The production of a capacitor in which two metal plates were separated by air proved that there could be no atmosphere extending out from the positively charged plate.[52] Another factor that argued against a possible role of these supposed atmospheres was the discovery (first reported in 1749, based on experiments of 1748) that negatively charged bodies, or "bodies having less than the common quantity of electricity, repel each other, as well as those that have more."[53] These, and other types of problems, led eventually to the abandonment of the doctrine of atmospheres and to other revisions of the theory.[54] But in the late 1740s the discoveries which set the seal on Franklin's scientific reputation were those connected with the Leyden jar.

Franklin must have been very proud to have received a letter from Musschenbroek telling him that "nobody has discovered more recondite mysteries of electricity than Franklin." He then added his hope that Franklin "would go on making experiments entirely on your own initiative and thereby pursue a path entirely different from that of the Europeans, for then you shall certainly find many other things which have been hidden to natural philosophers throughout the space of centuries."[55]

The Significance of the Lightning Experiments

Franklin's scientific fame was primarily produced by his experiments and his theory of electrical action, which—his admirers in England and in

France agreed—had made a science out of electricity. His *Experiments and Observations on Electricity* became one of the most widely reprinted scientific treatises of the Enlightenment.[56] His most spectacular achievement was to experiment with lightning. The lightning experiments gained for him a degree of international fame, in the minds of the scientific community and the public at large, that never would have been achieved on the basis of what we would call his fundamental laboratory investigations.

Franklin was not the first person to hypothesize an identity between the lightning discharge and the spark discharges produced by human operations. But Franklin went one step beyond all those who had noted similarities. He was actually able to design a test of whether clouds are electrified. He was able to make a five-word bold declaration, "Let the experiment be made!"[57]

Because Franklin's experiment made use of the properties of pointed conductors which he had discovered in the laboratory, no one before him could have designed this experimental test. He published a proposal to erect a kind of sentry box in which an experimenter might stand, protected from the rain, on an insulated stool. From the sentry box there could extend a long pointed rod which would "draw off" some of the charge from passing clouds.[58] This experiment was first performed in France, more or less according to Franklin's directions, on 10 May 1752. It was a great success. The experiment was repeated with equal success in England and elsewhere on the Continent.[59]

The significance of the lightning experiments is usually not fully understood. In order to appreciate what they meant to the scientific community, we must remember that in Franklin's day the subject of electricity was not yet fully recognized as a major or essential part of physics or natural philosophy. Not only was the subject characterized as a kind of "toy physics," accompanied by amusing parlor games for the entertainment of the curious, it also seemed to be a collection of phenomena that did not occur naturally but were the product of human artifice, or of human intervention in the processes of nature. Thus, electrical effects were produced by such "artificial" circumstances as rubbing a piece of sulphur or amber or a rod of glass, or causing the sphere or cylinder of an electrostatic generator to rotate while in contact with a "rubber."

Franklin's lightning experiments, showing that the lightning discharge is an electrical phenomenon, altered this situation in a radical manner. Every electrical scientist now knew that the experiments he performed with his little laboratory toys might reveal new aspects of one of the most dramatic of nature's catastrophic forces. The lightning experiments not only gave electrical studies a new importance in this sense; they also indicated that electrical phenomena occur in nature without human artifice or intervention. "The discoveries made in the summer of the year 1752 will make it memorable in the history of electricity," wrote William Watson, the leading figure in electrical science in Britain. These discoveries, he declared, "have opened a new field to philosophers." Even more important, the lightning experiments had finally shown scientists "that what they have learned before in their museums, they may apply with more propriety than they hitherto could have done, in illustrating the nature and effects of thunder, a phenomenon hitherto almost inaccessible to their inquiries."[60] From that day onward, no natural philosophy could be considered a complete or adequate account of nature if it did not include electrical phenomena along with those of mechanics, heat, optics, magnetism, pneumatics, hydrostatics, chemical interactions, and gravitational celestial mechanics. When, in the year following the lightning experiments, Franklin was awarded the Royal Society's Sir Godfrey Copley gold medal "for his discoveries in electricity," the president of the Society emphasized that not many years ago electricity "was thought to be of little importance, and was at that time only applied to illustrate in some degree the being and nature of attraction and repulsion." Indeed, until recently, nothing "worth much notice [was] expected to ensue from it." But now, he concluded, thanks to the ingenious discoveries and experiments of Franklin, electricity "appears to have a most surprising share of power in nature."[61]

It is well known that a by-product of Franklin's investigations of lightning was the invention of the lightning rod. The essential feature of the lightning experiments was the Franklinian discovery that an elevated pointed metal rod which was grounded could "draw off" some of the "electric fluid" from overhead clouds. So confident was Franklin of the anticipated result that he immediately recommended that rods be used in this way to prevent a lightning stroke. He assumed that such a rod would be able to rob passing clouds of their excess electric fluid, or at least some of it, so as to prevent or substantially reduce the chance of a stroke of lightning. Franklin published his design of a lightning rod in "Poor Richard's Almanack" even before the successful test.[62] In the event, the lightning rods proved to have an additional and even perhaps more effective action, insofar as they could attract an actual stroke and conduct it safely into the ground.

Franklin's lightning rod was important in two major ways. First of all, the lightning rod vindicated the prophecy of Bacon and Descartes that the pursuit of knowledge of nature must lead to practical inventions of real significance to mankind. Although this sentiment was echoed by many practicing scientists and advocates of science during the next century or more, there were no major instances of such fruits of the investigation into nature. Indeed, on a significant scale, there was no other major practical issue of pure science or fundamental research for perhaps another century. As late as the 1830s, in public debates in France over the eventual practical usefulness of pure scientific research, Franklin's invention of the lightning rod was still being cited as the primary example.[63]

Second, Franklin's work on lightning and his invention of the lightning rod took on a special value in an Age of Reason. All enlightened men and women were convinced that the advance of scientific discovery should result in the diminution of superstition and fear of natural phenomena. For centuries, lightning had typified the action of an angry father-God either sending a signal to his sinful children or hurling destruction upon them for their actions.[64] The Franklinian experiments showed that lightning is a naturally occurring phenomenon and therefore has no special divinely inspired quality.

Franklin's audacity in bringing lightning from the skies to the laboratory was later epitomized in the words of Turgot adapted from Manilius, "Eripuit coelo fulmen sceptrumque tyrannis," which was beautifully illustrated by Fragonard in an engraving entitled "Au génie de Franklin"[65]—"He snatched the lightning from the sky and the scepter from tyrants."

3

How Practical Was Franklin's Science?

In the minds of most people, Benjamin Franklin will ever personify the practical cast of the American mind. Many scholars and historians share this opinion. A typical expression of this point of view occurs in a statement made by Curtis P. Nettels, "Franklin, the most significant American of the eighteenth century and a utilitarian *par excellence,* directed his scientific thought to the improvement of material conditions."[1] The same writer also tells us:

> Franklin's early life as a printer—his practice of applying hand-work to ideas—explains in large measure the practical cast of his thought. No activity satisfied him so much as scientific inquiry—a mistress who did not have to be flattered and cajoled as did the human beings through whom he sought to accomplish his useful aims. His experiments that proved the identity of lightning and electricity established his international renown as a scientist. The invention of the Franklin stove (which sent the heat into the room instead of up the chimney), of the lightning rod, and of a clock having only three wheels and two pinions, all testify to the practical nature of his work.[2]

Any discussion of the practical or utilitarian character of Franklin's work in science must begin by making it clear that judgment should be made in terms of eighteenth-century values, not our own present-day values. In many ways the eighteenth century, like the seventeenth century, was somewhat more avowedly utilitarian-minded in statements about science than our own age. The influence of Bacon was strong, especially in England (and therefore in the American colonies), and the Baconian spirit embodied the goal of putting scientific discoveries to work for the improvement of man's material condition. No one would deny that Benjamin Franklin, a follower of the English school, was interested in putting his own scientific discoveries, *once they were made,* to some practical use. But this is greatly different from saying that he "directed his scientific thought to the improvement of material conditions."

Let us examine some examples to see what he actually said and did. A statement often quoted to illustrate Franklin's point of view was written as a supplement to a description of his discovery that the calorific effects of the sun's rays are absorbed in varying degrees by the same kind of cloth dyed in different colors:

> What signifies Philosophy that does not apply to some Use? May we not learn from hence, that black Clothes are not so fit to wear in a hot Sunny Climate or Season, as white ones; because in such Cloaths the Body is more heated by the Sun when we walk abroad, and are at the same time heated by the Exercise, which double Heat is apt to bring on putrid dangerous Fevers? That Soldiers and Seamen, who must march and labour in the Sun, should in the East or West Indies have an Uniform of white? . . .[3]

This document can be interpreted in one way only. It is an example of how Franklin, having made his discovery, thought of ways in which it could be put to use. He did not make the original experiments by "directing his thought" to any specific practical end. In fact, when he made the experiments, he could hardly have known that there would be any possible practical application, since at that time the fact of varying absorption of heat as a function of color was not even known to exist. It was discovered as a result of his experiments. Furthermore, the original experiments on which the proposed application was based had been made at least twenty-five years before the letter quoted above was written, and probably had been made as early as 1729, if not earlier. And yet the expression of a hope of possible application, at least according to the available writings of Franklin, did not occur until some three decades later, in 1761.[4]

It has also been alleged that "utilitarianism governed Franklin's conception of science—he once apologized for spending time in a mathematical exercise which could have no useful bearing."[5] The mathematical exercise in question is presumably that of making "magic squares" and "magic circles." The story of Franklin's interest in these mathematical curiosities is best told by himself:

> Being one day in the country, at the house of our common friend, the late learned Mr. *Logan,* he shewed me a folio *French* book, filled with magic squares, wrote, if I forget not, by one M. *Frenicle,* in which, he said, the author had discovered great ingenuity and dexterity in the management of numbers; and, though several other foreigners had distinguished themselves in the same way, he did not recollect that any one *Englishman* had done any thing of the kind remarkable.

> I said, it was, perhaps, a mark of the good sense of our *English* mathematicians, that they would not spend their time in things that were merely *difficiles nugae,* incapable of any useful application. He answered, that many of the arithmetical or mathematical questions, publickly proposed and answered in *England,* were equally trifling and useless. "Perhaps the considering and answering such questions," I replied, "may not be altogether useless, if it produces by practice an habitual readiness and exactness in mathematical disquisitions, which readiness may, on many occasions, be of real use." "In the same way," says he, "may the making of these squares be of use." I then confessed to him, that in my younger days, having once some leisure, (which I still think I might have employed more usefully) I had amused myself in making these kinds of magic squares, and, at length, had acquired such a knack at it, that I could fill the cells of any magic square, of reasonable size, with a series of numbers as fast as I could write them, disposed in such a manner, as that the sums of every row, horizontal, perpendicular, or diagonal, should be equal.[6]

Franklin continued to be fascinated by magic squares, and even by magic circles, despite his acknowledgment that they were useless and even a trivial waste of time. He corresponded about them and even consented to having one of them reproduced in the supplementary portion of the fourth English and third French editions of his book *Experiments and Observations on Electricity.*[7] And he wrote to his English friend and patron, Peter Collinson, "I am glad the perusal of the magical squares afforded you any amusement. I now send you the magical circle."[8]

One use that these magic squares might have, according to Franklin, was to provide a means of perfecting one's skill in arithmetic. In Franklin's own case, it seems to have worked. In his autobiography he tells us that at the "school for writing and arithmetic" which he attended until he was ten years old, "I failed in the arithmetic, and made no progress in it."[9] Yet the letter quoted above indicates that he had, in the end, acquired considerable arithmetical skill. Franklin's earlier ideas on this subject may be found in an article entitled "On the Usefulness of the Mathematics," published in the *Pennsylvania Gazette,* 30 October 1735.[10] It begins with the statement that arithmetic is valuable in "business, commerce, trade, or employment," even for "merchant and shopkeeper." As for geometry, "no curious art, or mechanic work, can either be invented, improved, or performed, without its assisting principles." This subject therefore is of great and indispensable use to astronomers, geographers, mariners, architects, engineers, surveyors, and so on.

But the "usefulness of the mathematics" is not limited to such obviously practical ends. We are told, for example, that "*Mathematical demonstrations* are a logic of as much or more use, than that commonly learned at schools, serving to a just formation of the mind, enlarging its capacity, and strengthening it so as to render the same capable of exact reasoning, and discerning truth from falsehood in all occurrences, even subjects not mathematical." "To give a man the character of universal learning, who is destitute of a competent knowledge of the mathematics" is, we are told, unreasonable or unjust.

In this instance, the fact of the matter is that Franklin made protestations of his lack of interest in an apparently non-useful subject, even to the point of expressing satisfaction that Englishmen had not bothered to waste time on such wills-o'-the-wisp as magic circles. In fact, the English mathematicians of Franklin's day were not simply Baconians, uninterested in things "incapable of any useful application." Nor were they following the precept of Robert Boyle: "The other humane studies I apply myself to are natural philosophy, the new mechanics and husbandry, according to the principles of our new philosophical colledge that value no knowledge but *as it has a tendency to use.*" Rather, it is one of the interesting facts of the history of mathematics that in contrast to the great period of the seventeenth century and that of the nineteenth and twentieth centuries, English mathematicians of the eighteenth century produced little of great value, especially as compared to what was produced on the Continent. And if Franklin seems to have apologized for so wasting his time, he nevertheless continued to spend a good bit of his leisure and interest on this useless subject. We must judge his scientific interests by his own actions, not merely quote again and again what he said about them.

Magic squares and circles appealed to Franklin's curiosity. They were a kind of game or puzzle. During the debates in the Assembly, he tells us, "I was at length [so] tired with sitting there . . . that I was induc'd to amuse myself with making magic squares or circles, or any thing to avoid weariness."[11] The pursuit of higher mathematics is in any case, according to the German mathematician David Hilbert, like playing a game in which one sets up the rules or operations and sees what results arise from the proper manipulation of the meaningless entities represented by the symbols.

The true test of the possible practicality of Franklin's scientific interest may come from the field in which he did his most significant work and which brought about his everlasting international fame, electricity. In discussing Franklin's electrical researches, most scholars lay great stress on his kite experiment and his invention of the lightning rod. Thus: "The invention of the lightning rod was only one illustration of Franklin's conviction that science should promote general well-being."[12] No one in his right mind would think for a moment of denying that Franklin's lightning rod was a practical invention. But it was an invention, and it must be classed along with other inventions like bifocal eyeglasses, the Franklin stove, and others.

There is only one difference between Franklin's invention of the lightning rod and his other inventions. The invention of the lightning rod was made possible by Franklin's own investigations into the phenomena of pure science. Franklin's interest in the subject of electricity was aroused by his contact with an itinerant lecturer on scientific subjects, Dr. Archibald Spencer, whom he met first in Boston in 1743 and whom he saw again when Spencer visited Philadelphia in the following year.[13] Soon after, he received a gift of some electrical apparatus with directions how to use it and began to experiment. Franklin wrote a letter to Collinson thanking him for the gift on 28 March, 1747, stating: "I never was before engaged in any study that so totally engrossed my attention and my time as this has lately done."[14]

Two letters describing the experiments and the conclusions to be drawn from them followed on 25 May, 1747 and 28 July, 1747.[15] A careful examination of these two letters, of some length, shows that they nowhere refer to any practical matter, or to any possible practical use that the discoveries being made in Philadelphia might have. Franklin was simply interested in a new subject, for which he found he had a considerable talent for investigation. As results were forthcoming, he continued his experiments with no motive other than that which impels any scientific investigator: he wanted to find the truth of the matter at hand, and that was all there was to it. On 14 August, 1747, after the two long letters just mentioned had been sent off to England, Franklin wrote a very interesting short note to Collinson, in which he admitted that some of his theoretical conclusions now appeared to have been a little hasty, that new experimental data seemed not to accord with his principles as originally laid down. He added: "In going on with these Experiments, how many pretty systems do we build, which we soon find ourselves oblig'd to destroy! If there is no other Use discover'd of Electricity, this, however, is something considerable, that it may *help to make a vain man humble.*"[16] This is not the kind of use that writers have had in mind when asserting that Franklin was interested in science only if it held promise of some application.

Another long letter to Collinson followed on 29 April, 1749. This letter contained much new material and ended on the following humorous note:

> Chagrined a little that we have been hitherto able to produce nothing in this way of use to mankind; and the hot weather coming on, when electrical experiments are not so agreeable, it is proposed to put an end to them for this season, somewhat humorously, in a party of pleasure, on the banks of

> [the] *Skuylkil.* Spirits, at the same time, are to be fired by a spark sent from side to side through the river, without any other conductor than the water; an experiment which we some time since performed, to the amazement of many. A turkey is to be killed for our dinner by the *electrical shock,* and roasted by the *electrical jack,* before a fire kindled by the *electrified bottle:* when the healths of all the famous electricians in *England, Holland, France,* and *Germany* are to be drank in *electrified bumpers,* under the discharge of guns from the *electrical battery.*[17]

Thus far, surely, despite Franklin's statement that he and his coworkers were "chagrined a little," nothing of any real practical use had been produced by the research.

Yet much of great value to the world of science had been found. Franklin had completed his analysis of the capacitor, or condenser, then known as the Leyden jar; he had formulated his great theory of a single fluid "plus or minus" or "positive or negative"; he had found the effect of shape on conductors, discovered the function of insulation and grounding in charging bodies, and unified all existing knowledge of this subject as well as made his own additions thereto. As yet, this subject was of no practical value outside the realms of knowledge. His work was acknowledged by the Royal Society as a significant contribution to our knowledge of nature and his findings were printed in a small book that was soon translated into French.

Franklin's experiments and observations had led him to notice the similarity between an electric spark and a flash of lightning and his own work showed him the way to test this hypothesis. This test took the form of the sentry-box experiment and was successfully carried to its conclusion under the direction of the French translator of his book and by others. As an afterthought, Franklin proposed the experiment of the kite and carried this through himself.[18]

The practical invention of the lightning rod followed directly from Franklin's own research. If lightning is electrical, its phenomena must accord with the laws of electrical action discerned in the laboratory. Thus, the electricity of clouds will be attracted by means of a pointed conductor (a new discovery). But the pointed conductor must be grounded (another new discovery). Finally, the electricity will choose to pass through a good metallic conductor rather than through the poor conductor of which the protected edifice is built (yet another new discovery). Only after such discoveries had been made could an experimental test be devised to see whether lightning is indeed an electrical phenomenon. Others before Franklin had guessed at the truth of this matter but could not prove it. There was not yet available the scientific basis for designing and executing an experimental test. These same discoveries laid the basis for Franklin's invention of the lightning rod. He thought at first that the clouds were positively charged and thus had an excess of electrical fluid which could be drawn out of them, thereby preventing a strike. As it turned out, the rod did not work in this way so much as to *conduct* a strike when it actually occurred.

The success of Franklin's invention of the lightning rod must have given him great pleasure in having produced something useful to humanity. Most scientists, of whatever century and in any field, experience the same pleasure whenever a discovery of theirs eventually leads to some useful application. But, in this case as in many others, no one could have predicted the practical result since no one knew at the beginning of the research where the experiments would lead. Franklin was interested in discovering scientific truths; he did not know beforehand that he would discover the nature of the condenser, the function of grounding and insulating, relative conductivities, the nature of pointed conductors—much less the electrical nature of lightning. The latter set the seal to his fame in the public mind. But his interest and enthusiasm were never dampened in the early days by the fact that he was simply discovering interesting and curious aspects of the world of nature.

Had "utilitarianism governed Franklin's conception of science," or had "a utilitarianism *par excellence* directed his scientific thought to the improvement of material conditions," he would never have begun to experiment with electricity at all. For this was a subject, when he began his investigations, of little, if any practical use. Only one application of electrical knowledge was known at that time, the therapeutic use of the electrical shock in certain cases of paralysis and other bodily disorders. Franklin himself did not believe in this type of treatment, although, like others, he did apply electric shocks for this purpose. He concluded that the supposed cures came less from the shock than from the patient's expectation of a cure.

Franklin could hardly have chosen a subject so apparently without promise of immediate practical application as the field of electricity. Of course, like others of his time, Franklin was deeply influenced by the Baconian philosophy and believed that any fundamental knowledge of nature would eventually be of some use to men and women in the control of their environment, that true science would necessarily produce useful inventions and practical applications. But this is a far cry from asserting that Franklin chose to engage in a research program in electricity because that was a subject especially likely to yield useful inventions as a by-product or because he envisaged any particular applications that might be produced.

One of the most practical results of Franklin's career in the field of electricity is usually not noted. It is that the spectacular success and fame arising from his scientific work made possible his successful diplomatic career. At the time of the Revolution, Franklin was one of the most distinguished living scientists.[19] His book on electricity had been published in ten editions in four languages. He was a member of the two most important scientific societies in the world, the Royal Society (London) and the Académie Royale des Sciences (Paris). Almost the entire scientific world accepted his theory of electricity and his book had become the vade mecum of electrical style; almost all who wrote on that subject used the terms which he had used in an electrical sense for the first time.[20] His book, in the words of a contemporary, "bids fair to be handed down to posterity as expressive of the true principles of electricity; just as the Newtonian philosophy is of the system of nature in general."[21] When Franklin went to the court of France, he was not an unknown man from an English colony; he was one of the best-known living scientists and benefactors of the age.

Franklin was interested in advancing knowledge, as all scientists are. In one of his communications on electricity, he wrote:

> These thoughts, my dear friend, are many of them crude and hasty; and if I were merely ambitious of acquiring some reputation in philosophy [i.e., natural philosophy, or science], I ought to keep them by me, till corrected and improved by time, and farther experience. But since even short hints and imperfect experiments in any new branch of science, being communicated, have oftentimes a good effect, in exciting the attention of the ingenious to the subject, and so become the occasion of more exact disquisition, and more compleat discoveries, you are at liberty to communicate this paper to whom you please: it being of more importance that knowledge should increase, than that your friend should be thought an accurate philosopher.[22]

Benjamin Franklin believed, as most scientists always have and still do, that all scientific discovery is eventually of some use—if not leading to immediate practical application, then at least useful in affecting the growth of some other branch of knowledge. When Franklin saw the first balloon ascent in Paris he overheard the remark "What good is it?" His reply, usually quoted as "What good is a newborn baby?" or even "What good is a newly born infant?" has been the standard reply of scientists and other innovators ever since.[23] It was used with particular effectiveness by Michael Faraday, who added, "The answer of the experimentalist would be, 'Endeavor to make it useful.' "[24] In inventing the lightning rod, Franklin endeavored to make his electrical discoveries useful.

In an age which was interested in the production of useful things, and in a cultural milieu which set great stock on being practical, Franklin nevertheless investigated scientific subjects primarily because they aroused his curiosity, because some particular topic excited his interest. When his investigations seemed far removed from the practical or useful sphere, he could not help having twinges of conscience about not being able to justify the expenditure of effort in terms of some practical return. Nevertheless, as a scientist, his interest was primarily in science itself and in scientific research for its own sake, and was never limited by considerations of only what might be useful.

4

The Mysterious "Dr. Spence"

Benjamin Franklin will always be remembered for his electrical discoveries, even should the future tend to denigrate or obscure his political accomplishments. Whenever we think or speak about electricity, we do so in the words introduced into the language of electricity for the first time by Franklin. The commonest electrical words—plus, minus, positive, negative, electric motor, electrical battery, and many others—were used in an electrical sense for the first time in Franklin's writings, and the popularity of his book *Experiments and Observations on Electricity* established them as standard for all time to come.[1]

We are quite naturally concerned to know how Franklin first learned about the subject in which his researches brought him such great fame. This topic is also of interest in providing an example of the ways in which scientific information migrated across the Atlantic to the American colonies. During the 1740s, when Franklin was introduced to electricity, he had been undergoing a thorough and rigorous course of self-education in the sciences.[2] Many of the books he had been reading contained discussions of the nascent science of electricity, as did periodicals such as the *Gentleman's Magazine*.[3] While these sources would have informed Franklin about some aspects of the subject, they evidently did not stimulate him to undertake research.

Franklin himself made three statements about the occasion of his becoming interested in electricity. Two of these occur in his autobiography, the third in a letter to Michael Collinson. According to the autobiography, Franklin saw his first electrical experiments in Boston. Here is his first account of this event and its consequences: "In 1746, being at Boston, I met there with a Dr. Spence, who was lately arrived from Scotland, and show'd me some electric experiments. They were imperfectly perform'd, as he was not very expert; but being on a subject quite new to me, they equally surpris'd and pleas'd me. Soon after my return to Philadelphia, our library company received from Mr. P. Collinson, Fellow of the Royal Society of London, a present of a glass tube, with some account of the use of it in making such experiments. I eagerly seized the opportunity of repeating what I had seen at Boston; and, by much practice, acquir'd great readiness in performing those, also, of which we had an account from England, adding a number of new ones."[4] A second statement in the autobiography contains another reference to Dr. Spence. This time we are told: "When I disengaged myself . . . from private business, I flattered myself that, by the sufficient tho' moderate fortune I had acquir'd, I had secured leisure during the rest of my life for philosophical studies and amusements. I purchas'd all Dr. Spence's apparatus, who had come from England to lecture here, and I proceeded in my electrical experiments with great alacrity."[5]

On the basis of these two autobiographical statements, scholars have generally assumed that Franklin met "Dr. Spence" in Boston in 1746, and that soon after his return to Philadelphia (1746 or perhaps early 1747) he and his fellow members of the Library Company received a glass tube from Peter Collinson in London with instructions concerning its use in electrical experiments.[6] Also, at some unspecified time, Franklin purchased the "apparatus" of Dr. Spence. One of the problems faced by scholars has been the total lack of information about a Dr. Spence other than Franklin's two references in the autobiography. Could the Doctor have passed through the colonies without a trace? Furthermore, the two statements by Franklin are contradictory. One says that he used Collinson's gift to repeat the experiments he had seen in Boston, and became quite adept in performing these and also additional ones (which he and his fellow experimenters had heard about from English publications). The other says that he purchased "all Dr. Spence's apparatus" and proceeded with his electrical experiments "with great alacrity." There is also a minor discrepancy about the origins of Dr. Spence: did he come from Scotland or from England?

Our perplexity is increased by a letter written by Franklin to Michael Collinson, the son of Franklin's patron and benefactor, Peter Collinson. The letter was an *éloge* of condolence on the occasion of the death of the elder Collinson. This letter reads in part: "Our dear departed Friend, Mr. Peter Collinson . . . transmitted to the Directors [of the Library Company] . . . every philosophical Discovery; among which, in 1745, he sent over an account of the new German Experiments in Electricity, together with a Glass Tube, and some Directions for using it, so as to repeat those Experiments. This was the first Notice I had of that curious subject, which I afterwards prosecuted with some Diligence, being encouraged by the friendly Reception he gave to the Letter I wrote him upon it."[7] According to this letter, Collinson's gift was "the first Notice" which Franklin had of "that curious Subject" electricity, whereas in the autobiography he saw the experiments of Dr. Spence in Boston before the receipt of Collinson's gift. The portion of the letter just quoted also contradicts the autobiography with regard to the date of Collinson's gift. The letter to Michael Collinson would place the date of this gift at 1745, while the first extract quoted above from the autobiography would place it at 1746 (or even 1747).

Hence, the historian is faced with a number of questions. Did Franklin see the demonstrations of Dr. Spence before he received Collinson's gift? Did Franklin encounter electricity for the first time in 1745 or in 1746? Or even earlier? Who was Dr. Spence? What did he actually demonstrate? What was the content of his lectures? When did Franklin purchase his apparatus? What were the "directions" for using the "Glass Tube" sent by Collinson? And what was the "account of the new German Experiments in Electricity"?

In the solution of the puzzles it will be seen that Franklin's "a Dr. Spence" was apparently Dr. A. Spencer (in fact, Archibald Spencer, M.D.). Franklin first encountered him in Boston in 1743 and later in Philadelphia in 1744. Earlier attempts to identify Franklin's mysterious Dr. Spence failed because of the wrong date and name. Franklin thus did encounter Spencer before receiving Collinson's gift in 1746 and in fact he acted as Spencer's agent for the latter's lectures in Philadelphia in 1744. But he quite evidently did not purchase Spencer's apparatus until some few years later.[8]

Announcements of Spencer's Lectures

An advertisement which appeared in the *Boston Evening Post,* no. 408, 30 May 1743, contained an announcement that "Dr. Spencer having a compleat Apparatus, proposes to begin a Course of Experimental Philosophy in Boston." This advertisement was repeated in the issue of 6 June 1743 and also appeared in the *Boston Weekly Post-Boy* no. 441, the issue of 30 May 1743; another advertisement appeared in the *Boston Evening Post,* no. 408, the issue of 1 August 1743:

> Doctor Spencer having a compleat Apparatus, proposes to begin a Course of *Experimental Philosophy* in Boston, as soon as Twenty shall have subscribed (of which Notice shall be given) to be continued at such Times as shall be agreed upon by the Subscribers at the first Lecture. The charge of going through the Course is *Six Pounds,* Old Tenor, to be paid one Half at Subscribing. Those that are inclined to attend, are desired to enter their

Names, and pay the Subscription Money to Mr. *Thomas Kilby* at the Naval Office, who will furnish a Catalogue of the Experiments, gratis.[9]

According to the terms announced by Spencer, the "Course" would be given only if twenty subscribers would enroll. In that event, Dr. Spencer declared, the time and place of the first meeting would be published in the newspapers. Since I have not been able to find such an announcement, I conclude that the lectures were not given. Perhaps the cost, "Six Pounds, Old Tenor," may have been too great a price for the Bostonians to pay for Spencer's scientific lectures.

Some time later, Spencer came to Philadelphia, according to an advertisement which appeared in Franklin's *Pennsylvania Gazette,* no. 802, 26 April 1744:

> A Greater Number of Gentlemen having subscribed to *Dr. Spencer's* first Course of EXPERIMENTAL PHILOSOPHY, than can be conveniently accommodated at a Time: He begins his first Lecture of the second Course, on Thursday, the tenth day of May, at five o'Clock: Subscriptions are taken in at the Post Office, where a Catalogue of the Experiments may be had *gratis.*

Franklin was at that time Postmaster of Philadelphia and the post office was in his house on Market Street. Since subscriptions to the lectures were "taken in at the Post Office," where one could obtain a "Catalogue of the Experiments," we can be certain that Franklin knew of these lectures. In effect, he acted as Spencer's agent. No copy of this "Catalogue of the Experiments" has been discovered or identified.

Franklin frequently referred to his business establishment simply as "the Post Office." For example, in the *Pennsylvania Gazette,* no. 798, 29 March 1744, Franklin published the following notice:

> A COLLECTION of choice and valuable BOOKS, consisting of near 600 Volumes in most Faculties and Sciences, viz. *Divinity, History, Law, Mathematicks, Philosophy, Physick, Poetry,* &c. Will begin to be sold by Benjamin Franklin, for ready Money only, on Wednesday the 11th of April 1744 at Nine o'Clock in the Morning, at the Post-Office in Philadelphia; the lowest Price being for Dispatch marked in each Book: Catalogues may be had *gratis,* at the Place of Sale.
>
> Note. *The said Franklin, gives ready Money for any Library or Parcel of Books.*

Spencer's announcement, quoted above, was reprinted in the *Pennsylvania Gazette* of the week following—no. 803, 3 May 1744, the following advertisement:

> A COURSE OF EXPERIMENTAL PHILOSOPHY, begins at the Library-Room, next Monday at five o'Clock in the Afternoon, which will be the last to be performed in this City by Dr. *Spencer.*

N.B. Any of the Gentlemen who subscribed to the former Courses, may go through this, at half Price, and have as an Addition some *Lectures* on the Globes.

Thus we know that Spencer offered a course in experimental philosophy in Boston in 1743, but apparently did not obtain a sufficient number of subscribers to warrant his beginning the lectures. He certainly came to Philadelphia in the spring of 1744 and attained so high a degree of success that he had to repeat his course several times in order to satisfy the demand. In Philadelphia, Franklin acted as his agent. Knowing Franklin, we may be certain that he would have attended Spencer's lectures and demonstrations in Philadelphia in 1744 and would have discussed certain scientific subjects with him. We shall see, below, that in an early draft outline of the autobiography, Franklin assigned the date of 1743 to his first encounter with "Dr. Spence" in Boston.

The Content of Spencer's Lectures

In the course of my research on Spencer's course of lectures, I was fortunate in finding two sets of notes taken by men who had been present. One such source is the journal (1744) kept by William Black, who wrote out two accounts of Spencer's lectures. The first of these, entered under the date of 29 May 1744, reads:

> Between the hours of 3 & 4 the Governor, Commissioners, and the rest of the Company went to hear a Philosophical Lecture on the Eye, &c. by A: Spencer, M:D:, in which he endeavoured to account for the Faculties, the Nature and Diseases of that Instrument of Sight; next he proceeded to show that Fire is Diffus'd through all Space, and may be produced from all Bodies, Sparks of Fire Emitted from the Face and Hands of a Boy Suspended Horizontally, by only rubbing a Glass Tube at his Feet.[10]

The second entry appears under the date of 5 June 1744:

> At 11 in the forenoon, with Colonel Beverley and the Gentlemen of the Levee, I went to the State House, where Doctor Spencer Entertain'd Us very Agreeably with several Philosophical Transactions, first he Prov'd and Illustrated by Experiments, Sir Isaac Newton's Theory of Light and Colours, also Several Curious Objects Shown by the Solar Microscope, together with the Circulation of the Blood, all he perform'd very much to the Satisfaction of the Spectators.[11]

These brief notes can be considerably amplified by a second set which I found in manuscript in the Library Company of Philadelphia, among the papers of John Smith.[12]

Smith's notes are not labeled and there is no direct evidence that they were written on hearing Spencer; that is, they do not mention Spencer's name, nor are they dated. But, since Smith was in Philadelphia in 1744, and since he was interested in science, we can reasonably assume that he attended Spencer's lectures. The similarity between his notes and those taken by William Black surely provides ample evidence that both were derived from the same lectures. Smith's notes read as follows:

Lecture 1st.

That the properties of all matter is the same Viz. Solidity, Extention, Divisibility, mobility & passiveness or Inactivity—that upon this principle the Softest body is as Solid as the hardest, And the Lightest as the heaviest. The Matter of Cork is as Solid as Lead, nor is a Diamond a Jot more Solid than Water. Solidity is a property of matter whereby it Excludes every other body from the place itself possesses. Extension is that whereby a thing is Constituted Long, Broad, or thick &Ca.:

Divisibility, a passive power, or property in quantity whereby it becomes separable into parts, either actually or at least mentally. The Dr. Supposes, Matter admits of a Mathematical Divisibility ad Infinitum—as in this Example 1/4 one divided into four parts. If I add an 0 it [the denominator] is 40 if another it is 400 if another 4000, & so endless. Upon this head he observed that One Grain of Gold, by the Workmens Coarse Tools, could be workt into 50 square Inches—that the Length of an Inch could be Divided into 200 Visible parts, so that in one Square Inch there is 40 000 Visible parts; & in one Grain of Gold there are 2 000 000 of such parts; which Visible parts no one will deny to be farther Divisible. Again, an ounce of Silver may be Gilt with 8 Grains of Gold, which is afterwards drawn into a wire 13,000 foot Long. By the help of Glasses he said One may See Many ridges or Scales on the Surface of the hand, that by the Microscope one may Count 500 pores in one of those Scales & one grain of Sand will Cover 200 Scales—By the microscope many small Animals may be perceived A thousand of which (says the Dr.) may lead up a Country Dance on the point of a needle. And yet these must have all the parts necessary for Life, as blood, & other Liquors: How Wonderful then must be the Subtilty of the parts which make up such fluids!

Mobility is an aptitude or facility to be moved.

Passiveness or Inactivity is that power whereby Matter Resists any Change Endeavoured to be made in its State. I.E. whereby it becomes difficult to alter its state, either of rest or motion. So that it requires the same force to stop any thing when in motion, as to put it into motion.

Upon Gravitation the Dr. Observed, that Every Body Inclines towards a Center. That all Bodies near the Earth, have a Gravity or tendency toward the Centre of the Earth—. That this power Encreases as we descend & decreases as we Ascend from the Centre of the Earth, and that in the proportion of the Squares of the Distances therefrom reciprocally; so as, for Instance, at a double distance, to have but a quarter of the force &Ca.

Two Leaden Balls, almost square & Rubbed together, stuck so close that it required a great Weight to Separate them.

Divide a Loadstone in two, & one will attract to the North; the other to the South, also two needle points being rubbed on the same Loadstone & put in Water, the Loadstone attracted one and repelled the other.

*****

One of the Drs Experiments relating to Electrical Attraction was thus—

He took a Long Glass Tube, & Rubbed it Vehemently with his hand, & than held it pretty near several Pieces of Leaf brass or Gold, which put them into very brisk & Surprizing Motions. Some would leap toward the Tube, Sometimes adhere & fasten to it, settle on its Surface, and there remain Quiet: and sometimes be thrown off from it with a great force. And thus would they be alternately attracted and Repelled, for several times Successively. Sometimes, again, they would move slowly toward the tube, sometimes, would remain suspended between the tube and the Block they were first laid on; and sometimes slide along in the direction of the side of the tube, without touching it.

The Dr. Argued for a Vacuum thus. Supposing the particles of Air to be Globular. Suppose them to be what shape you please. If you Admit that the Air is Composed of particles—there must be some distance between each. Consequently a Vacuum. Again. If a small parcel of Air can be Rarified so as to Occupy a Larger Space than it naturally would do, the Vacuitys or distance between the particles must be greater. That the Air is Compounded of Particles you must allow because it has the properties of Matter as Solidity; Mobility &Ca. and you cannot Suppose Matter without particles.

A Glass of Malt Beer being put under a Receiver & the Air Exhausted, the Beer immediately set to working, & after working a while being taken out it tasted Flat & Dead.

A little Cool Air is proper in a fever but no Cordials.

All Bodies to which fire are applyed, become bigger, Swell'd and Rarifyed.—Spirit contracts in cold weather a 50th part & Expands in hot.

A Glass Ball with Rum & one with Water made thus [sketch] the Ball being put into hot Water the Rum rose in the Tube much faster & higher than the water.

A Tin Vessel filled with Oyl over a fire will melt before the Oyl will boyl.

Large Bodies have always a lesser Surface in proportion to their matter than small ones.

Organs of Sence Uncertain Judges of Heat. E. Gr., That Air after a Storm in August which we Suppose Cool is really Warm.

Heat puts the Particles in Perpetual Motion. Cold Inclines them to rest & Quiet. A Globe with a Retorted pipe the end of the pipe being put in Cold Water & the Globe in hot (such a one as this [sketch]) The heat Rarifying the Inclosed Air caused it to bubble out thro' the water, & the Globe being taken out of the hot Water, the Cold arose in the Pipe.

An Experiment was made on the Pyrometer of the different degrees of

Expansion of a Bar of Copper & a Bar of Iron by the same heat, & both being heated 40 degrees, the Iron Expanded 24 and the Copper 56.

*****

Galileo Discovered the Elasticity or Spring of the Air & proved it by this Experiment. An Extraordinary Quantity of Air, being Intruded by means of a Syringe into a Glass or Metal Ball. Upon taking the Syringe away, the Air rushes out. Consequently it is Endued with an Elastic force.

Bubbles are dilatable or Compressible. I. E., take up more or less room as the Included Air is more or less heated, or more or less pressed from without; and are round because the Included Aura acts Equably from within all around. Their Coat or Cover is formed of the minute particles of the fluid, retained either by the Velocity of the Air, or by the brisk Attraction between those minute parts and the Air.

The Lower Air is always denser than the Upper: Yet the Density of the Lower Air is not proportional to the Weight of the Atmosphere by Reason of Heat & Cold which make notable Alterations as to rarity & Density.

*****

The Light is only seven minutes passing from the Sun to us.

The Globe of the Eye Consists of 6 Coats & 3 humours. For an Inflammation in the Eye, Cupping by the side of it is Good.

All our Senses are feeling.

Oyl of Vitriol & Cold Water mixt became Immediately Warm.

Oyl of Vitriol & Oyl of Tartar mixt produced Ebulition.

Spirit of Nitre & the Oyl of Cloves mixt will produce flame.

A Bullet out of a Gun moves 1142 feet in a Second & the Strongest Wind but 35 foot & says the Dr. its friction against such a quantity of particles of Air, heats it as it goes, so that when it falls it is very warm, & the further it goes, the hotter it is.

The Strongest Winds are always attended with a hot Air.

Take the filings of Steel & Sulphur, Tempered with Water, & put under Ground will produce an Earthquake.

Clouds of Ice melting together in the Air by their Violent Attrition may cause a Lightning, & this heat Rarifying the Air, Other Air rushing in there causes Thunder.

Fire is diffused thro' all Space, contained in, & may be produced from all bodies.

A Boy was suspended Horizontally & the Dr. rubbed a Glass Tube, a little distance from his feet which made Sparks of fire fly from his face & hands.[13]

Before analyzing the contents of these lectures and discussing their significance for the development of Franklin's scientific thought, I shall present some further information about Spencer and Franklin's contact with him.

Additional Information about Spencer

Although we have no record that Archibald Spencer lectured in Philadelphia after the July series of 1744, we know that he was still in Philadelphia during September. In the "Itinerarium" of Dr. Alexander Hamilton we find the following extract under the year 1744: "Friday, September 14th.—I . . . drank tea with my landlady Mrs. Cume, and at five o'clock went to the coffee-house, where I saw Dr. Spencer, who for some time had held a course of physical lectures of the experimental kind here and at New York. . . . I met here likewise one Mitchell, a practitioner of physick in Virginia, who was travelling as he told me upon account of his health."[14] The Mitchell in question is Dr. John Mitchell, for many years resident in the state of Virginia and famous for his botanical activities. He returned to England in 1746 and soon afterward published his famed *Map of the British and French Dominions in North America* (London, 1755).[15]

Sometime between the spring of 1743 when he was in Boston and the spring of 1744 when he was in Philadelphia, Spencer visited New York. In a letter dated 22 January 1744, James Alexander wrote to Cadwallader Colden:[16] "The Comet was Seen at Philadelphia before Christmas—I heard nothing of it till the 2d of January when Dr Spencer told me that he was told a blazeing Star was Seen by the people. I had askt him & Mr Kennedy to Spend the Evening, but was kept from them by Alderman Johnson Mr Murray & Mr Smith . . . , when they departed casting up my Eyes to the Stars at first sight I saw the Comet & went in & told Mr Kennedy & Dr Spencer I had seen it & it was very visible—as I had their Company I Did not Attempt to make any Observation of it that night."[17] This is the same comet concerning which Franklin informed his readers throughout the winter at the end of 1743 and the beginning of 1744. One such notice, typical of the set, appeared in the *Pennsylvania Gazette,* no. 792, 16 February 1744, sandwiched in between news of ship arrivals in Philadelphia and a notice concerning "the Body of one Smith" which "was found floating in the River near Point Nopoint." This notice reads: "The Comet, which has been seen in the Evening these Six Weeks past, now appears in the Morning, rising before the Sun. It is very much encreased in Bigness and Length of Tail since its first Appearance."

Approximately one year after Hamilton recorded his meeting with Spencer and Mitchell, the latter wrote a letter to Colden which is of double interest. It contains news of Spencer and it throws light on Franklin's early unsuccessful attempts to start a scientific society in Philadelphia. This letter is written from Mitchell's estate in Urbanna, Virginia, and is dated 10 September 1745. Concerning an essay of his, he writes: "I thought indeed that it might make a fit piece for the records of the

Philosophic Society at Philadelphia; but was sorry to find none such, . . . & am still more sorry to hear, by Dr Spencer, that their laudable design is in a manner entirely dropt."[18] Colden replied to Mitchell in a letter dated New York, 7 November 1745, which concluded with the words "Please . . . to give my Service to Dr. Spencer."[19]

A few further items concerning Spencer's career were uncovered by Carl Bridenbaugh in the course of his research into eighteenth-century Philadelphia culture. One of these occurs in Thomas Cadwallader's *Essay on the West India Dry-Gripes,* published by Franklin in Philadelphia in 1745. Bridenbaugh informs me that there are two copies in the College of Physicians in Philadelphia, with different prefaces. In the second, Cadwallader thanks "my learned and worthy Friend, Dr. A. Spencer, of Philadelphia, (who is justly recommended by the famous Dr. Mead, and several eminent Gentlemen of the Faculty in London, as a most judicious and experienced Physician and Man-Midwife) for his Trouble in revising this Essay . . ." Carl and Jessica Bridenbaugh also found out that Spencer had been trained in Edinburgh and went to Newport some time between the summer of 1743 (when he was in Boston) and the spring of 1744 (when he was in Philadelphia). In Newport he encouraged Dr. Thomas Moffatt and the ingenious clockmaker William Claggett to undertake electrical experiments.[20]

Was Archibald Spencer Franklin's "Dr. Spence"?

As a result of the researches of J. A. Leo Lemay and R. P. Stearns, subsequent to my original findings, we now know that Archibald Spencer lectured along the Atlantic seaboard until at least 1751, when he was ordained and accepted a living in Maryland.[21] There is some uncertainty about when Franklin purchased Spencer's apparatus (as he recorded in his autobiography). The date of purchase must have been quite some time after Spencer's Philadelphia lectures of 1744, however, since Spencer continued to give lectures for several years afterward. Announcements of Spencer's lectures appeared in the *Virginia Gazette,* 9 January 1745/46 (a lecture in Williamsburg), and in the *Maryland Gazette,* 26 September 1750 (a lecture in Annapolis).[22] Since Spencer continued to lecture on experimental science after leaving Philadelphia, Lemay has concluded that Spencer would not have sold his apparatus "while on a lecture tour." The sale would probably have occurred "after Spencer was ordained and had a living (All Hallows, Anne Arundel County, Maryland) in the fall of 1751." The probate records indicate that Spencer possessed no scientific or lecture apparatus at the time of his death in 1760. Lemay therefore concludes that Franklin may not have purchased Spencer's apparatus "until January 1754, when he visited Annapolis."[23]

Some additional information on Spencer is found in a letter from John Bartram to Cadwallader Colden: "the next fifth night we are to have another meeting where Dr. Spence will accompany us. He exhibits Philosophical lectures now at Philadelphia & approves of our design: offers to take our proposals with him to the west indies with favourable account of our proceedings."[24] The "we" must refer to the planned philosophical society and "our proposals" are apparently those printed by Franklin.

We now come to the question: Can we be certain that Archibald Spencer, M.D., is the mysterious "Dr. Spence" of Franklin's autobiography? Franklin certainly knew of Spencer, since he published advertisements concerning his lectures in the *Pennsylvania Gazette,* acted as his agent to the extent of distributing the programs of the lectures, and was a member of the board of directors of the Library Company of Philadelphia (in whose rooms in the State House Spencer gave his course). In order to use the rooms of the Library Company, one first had to obtain the permission of the directors. And as if that were not sufficient, Spencer was acquainted with at least four of Franklin's friends and scientific correspondents—Bartram, Colden, Alexander, Mitchell. There can be no doubt that Franklin knew of Spencer's activities in Philadelphia.

Anyone who knows anything at all about Benjamin Franklin would find it hard to believe that a man of his curiosity and inquisitiveness and with his great interest in science would have limited his relations with Spencer to the purely commercial. He would certainly have talked with him about scientific questions and would have attended his lectures as well. But even if it were to turn out that Dr. Archibald Spencer is not "Dr. Spence" we must, in any case, revise our notions concerning the date of Franklin's introduction to electricity, since Spencer was in Philadelphia in 1744 and supposedly Franklin did not encounter "Dr. Spence" until 1746.

Except for the date and the final *r* to the name, the data concerning Spencer and "Spence" are identical, as may be seen as follows:

1. According to the autobiography, a "Dr. Spence" was in Boston in 1746 where Franklin met him. Again according to the autobiography, Spence showed Franklin some electric experiments. Franklin writes specifically that he "show'd me some electric experiments," from which we may assume that Franklin was given a private showing and did not attend a public lecture-demonstration.

According to the newspapers, Spencer was in Boston in 1743 and offered to give a course of lectures on experimental philosophy, but apparently could not obtain a sufficient number of subscribers to warrant giving the course.

While Franklin's statements in the autobiography may raise doubt concerning the date when he actually met "Dr. Spence," the "Draft Scheme" of that work states decisively that he did meet the doctor in Boston in 1743. Here one finds two significant successive phrases, which I have italicized in the following excerpt: "Success in Business. Fire Companies. Engines. *Go again to Boston in 1743. See Dr. Spence.* Whitefield. My connection with him. His generosity to me. My return . . ."[25] We may rest assured, I believe, that Franklin first encountered Spencer in Boston in 1743.

As Carl Van Doren has documented, Franklin visited New England during the spring of 1743, that is, during the time when Spencer was making his unsuccessful effort to obtain members for the projected course.[26]

2. Later, "Dr. Spence" came to Philadelphia "to lecture" and Franklin bought "all Dr. Spence's apparatus."

In 1744, Spencer came to Philadelphia and lectured over a period of several months.

As I have pointed out earlier, Franklin's reliability on the question of the date of his introduction to electricity is questionable. He wrote that Peter Collinson's gift of an electric tube arrived soon after his return to Philadelphia from the Boston visit (said by Franklin to have occurred in 1746) during which he saw "Spence." The autobiography thus places the date of Collinson's gift at 1746 or 1747. But, in Franklin's letter to Collinson's son Michael, the date of this gift is placed at 1745. If we knew the exact date of Collinson's gift, we might have an immediate answer.

The gift was sent to the Library Company but, unfortunately, the Minute Book is very incomplete for the early period and contains no record of Collinson's gift. Both the autobiography and the letter to Michael Collinson were written many years after Franklin had begun his work in electricity, and he may not have remembered exactly at what date before 1747 (when he wrote his first letter on electricity to Collinson) he had received a gift of an electric tube.[27] He would have recalled that some time before the gift arrived he had seen Spencer's demonstration (or "Spence's"). And if he did not remember the date correctly, he may just as well have forgotten that the name was actually "Spencer" and have thought it was "Spence."

Franklin was in close contact with Spencer, who can readily satisfy the account of "Spence" given in the autobiography if we are willing to accept two very plausible errors on Franklin's part. If we are not willing to accept the identification of "Dr. Spence" as Dr. Archibald Spencer, then we must make the unlikely supposition that two men, with almost identical names, came to America within a period of two years, that both went first to Boston (and did not give public lectures) and then to Philadelphia (where they did give public lectures), that both knew Franklin, and that both gave lectures on electricity, but that whereas one left behind him an easily traceable record, the other left behind no trace of his visit save a mention in Franklin's autobiography.

Even apart from the fact that it is more reasonable to suppose that Dr. Spencer is the true "Dr. Spence," there is a last bit of evidence to support this proposition. From the notes of Spencer's Philadelphia lectures, we learn that he demonstrated the circulation of the blood. Since the notes would not appear to indicate that the doctor performed a dissection, one would suppose that he accomplished his demonstration by means of a machine designed to show the action of the heart and circulatory system, a common enough device in the eighteenth century. Franklin, in his autobiography, does not say that he bought only the Doctor's electrical apparatus, but rather than he bought "all" of "Dr. Spence's apparatus, who had come to lecture here." Hence, he would have bought Spencer's apparatus for demonstrating the circulation of the blood.

Franklin, in fact, owned such an apparatus, which he treasured as one of his prize possessions. When Manasseh Cutler visited the aged Franklin, he reported that he was shown a "glass machine for exhibiting the circulation of the blood in the arteries and veins of the human body."[28] This would not appear to have been the makeshift device which Franklin in 1745 described to Colden as "a siphon made of two large joints of Carolina cane . . . into which two small glass tubes are to be inserted."[29] From Cutler's account, it would seem that what he saw was a solid or permanent piece of apparatus, and not such a rude contrivance. After 1745, Franklin never again returned to the subject of the circulation of the blood and it seems unlikely that he would have kept his temporary device until he died. On the other hand, a permanent well-made apparatus of glass, such as he bought from the doctor, was the sort of instrument which he would have kept proudly in a cabinet of scientific curiosities.

To sum up briefly, the accumulated evidence would tend to show that in its form as printed, Franklin's autobiography contains two errors concerning his introduction to electricity. First, the person who showed Franklin his initial demonstration of electrical phenomena was not "a Dr. Spence," but rather Dr. A. Spencer. Second, rather than in 1745,

Franklin's meetings with the doctor took place in Boston in the spring of 1743 and in Philadelphia in the spring and summer of 1744.

What Did Franklin Learn from Spencer's Lectures?

Dr. Spencer evidently dispersed a few medical precepts among his remarks on natural philosophy. Remarks such as "A little Cool Air is proper in a fever but no Cordials," and "For an Inflammation of the Eye, Cupping by the side of it is good," served to alleviate the barrenness of pure science by a little bit of practical advice for the benefit of the audience.

As a whole, the notes made by Smith and Black are interesting in displaying the kind of science which was being diffused in the Colonies in the 1740s. There was a demonstration of Newton's discovery of dispersion and the nature of color, of Harvey's discovery of the circulation of the blood. We may note, too, that Spencer referred frequently to experiments and to measurements (that is, to numerical data) and that he performed many experimental demonstrations to illustrate the topics on which he lectured.

Our attention centers quite naturally on Spencer's remarks concerning electricity. Following a discussion of the principles of solidity and infinite divisibility of matter, and of mobility and passivity, he turned to gravitational attraction and then to magnetism. He showed his audience that if a lodestone is broken, the newly exposed ends of the two pieces exhibit opposite polarities. Spencer next addressed himself to electricity, a subject related to magnetism in that both deal with attraction and repulsion.

He showed his audience first that if a glass tube is rubbed, it will attract small pieces of metal, and that some of these, after being attracted so as to touch the rubbed glass, will then be repelled. Electrical attraction of this kind (created by rubbing amber) had been known to the Greeks; repulsion had been discovered in the seventeenth century by Cabeo.[30] These simple phenomena represent knowledge on such an elementary level that Franklin, in his book on electricity, took them for granted and did not bother to describe them, although until Franklin put forth his unified single-fluid theory of electrical action they were not well explained.

The concluding item in Black's account as well as in Smith's concerns the electrification of a boy suspended horizontally from the ceiling. This experiment had been performed for the first time by Stephen Gray and announced by him on 8 April 1730.[31] Electricity became a popular subject in the middle of the eighteenth century and, of all the experiments

Frontispiece to Nollet's *Essai sur l'électricité des corps* (Paris, 1746). Houghton Library, Harvard University.

used to demonstrate the novelty of the subject, this one was certainly one of the most widely used.

Franklin himself never performed this particular demonstration so far as we know. But he did electrify human beings for more genuinely scientific purposes, such as to test the value of an electric shock as a therapeutic agent in certain types of paralysis, and he frequently electrified himself and others in the course of various electrostatic experiments. Rather than suspend an innocent victim from the ceiling by means of silk threads, Franklin made use of the equally effective if less dramatic method of having the person to be electrified stand or sit on a glass stool, which insulated the person equally well.

Two years after Spencer lectured in Philadelphia, the Leyden jar was invented and the whole science of electricity was speedily transformed. The new invention offered a means of "storing" an electric charge and opened up the possibility of experimenting with electricity on a large scale. Franklin's work was accomplished on a higher level of electrical knowledge than that demonstrated by Spencer in Philadelphia. The only immediate link between Franklin's writing on electricity and Spencer's lectures occurs in the penultimate sentence in Smith's notes. "Fire is diffused thro' all Space, Contained in, & may be produced from all bodies." By "fire" Spencer referred to electricity, which was then often called "electrical fire" or simply "fire."

N. V. de Heathcote has called attention to the fact that both Black's and Smith's notes describe how Spencer showed that "fire is diffused through all space" (the same wording occurs in both sets of notes); he "suspended" "a boy" "horizontally" and "made sparks of fire fly from his face and hands" ("sparks of Fire Emitted from the Face and Hands of a Boy," by "only rubbing a Glass Tube at his Feet.")[32] In these two descriptions, according to Heathcote's interpretation, the experiment is described not as an electrical manifestation but "as an instance of the universality of 'fire' . . . Neither Black's nor Smith's notes on the suspended boy experiment contain any suggestion that the experiment was of an electrical nature."[33]

Even though Smith's notes do contain descriptions of another set of experiments under the heading "Electrical Attraction," the words "electricity" and "electrical" do not appear in the text. Additionally Heathcote found that the text itself follows very closely the wording of Hauksbee's celebrated and well-known *Physico-Mechanical Experiments*.

Heathcote concludes that the experiments with a rubbed glass tube and brass leaf that Franklin saw being performed by Spencer in Boston in 1743 and in Philadelphia in 1744 "naturally interested him," but may have "meant little to him at the time." Hence there arises the possibility that it was only after the Library Company received Collinson's gift of an electric tube that Franklin would have "realized for the first time the true character of the experiments he had seen attempted three years earlier." Heathcote accordingly suggests that it could thus have been correct for Franklin to have said that "the first time he saw such experiments was when he was in Boston" (in 1743), but that the "first Notice" he had "had of electricity was from Peter Collinson in 1746." Heathcote thus suggests in conclusion that "Spencer introduced Franklin to an experiment with a glass tube and leaf-brass; Collinson introduced him to *electricity*."[34]

More recently, John L. Heilbron has argued that even though the "surviving accounts of [Spencer's] lectures . . . do not explicitly mention electricity," Spencer, who appears to have been competent and well informed, no doubt explained that "the fire drawn from the dangling urchin was electrical in nature." Heilbron considers it "unlikely" that "Spencer did not know, or failed to advertise, Gray's discovery of the connection between sparks and electrical attractions." Rejecting Heathcote's reconstruction, Heilbron finds it "more plausible" that Franklin's interest in electricity was not stimulated by Spencer primarily because "Spencer's repertoire was too poor in electricity to intrigue Franklin." In particular, he calls attention to the fact that "Spencer had left Europe before he could have learned of the fresh electrical games of the German professors."[35] I find Heilbron's argument, especially the reference to the recent electrical games or amusements, very convincing.

Two years after Franklin began the investigation of electrical phenomena he wrote his paper entitled "Opinions and Conjectures concerning the Properties and Effects of the Electrical Matter . . . Made at Philadelphia, 1749." In this paper, Franklin, for the first time, used the expression "electrical fluid" for electricity, abandoning the older expression "electrical fire." The opening paragraph of this paper of Franklin's still uses the term "electrical matter": "The electrical matter consists of particles extremely subtile, since it can permeate common matter, even the densest metals, with such ease and freedom as not to receive any perceptible resistance."[36] This may be compared with Spencer's statement that electrical "fire" is "diffused thro' all Space, Contained in, & may be produced from all bodies."

That Franklin and Spencer should be in accord is no occasion for surprise. Like all other sons of the eighteenth century, both were thinking in terms of an "imponderable fluid" which could pass in and out of matter with little or no resistance, that is, a "subtile" fluid. What distinguishes Franklin's theory of electricity from those of his contemporaries and predecessors is not that he conceived of a subtle fluid of electricity, but rather the particular use he made of such a fluid in establishing a unitary theory of electrical action.[37]

Scientific Lectures in Mid-Eighteenth-Century Philadelphia

Dr. A. Spencer was far from being the only lecturer on scientific subjects to visit Philadelphia during the 1740s. And Franklin's activities on his behalf were hardly a unique service. Several years prior to Spencer's visit, Franklin was sponsor and agent for another visiting lecturer on science, Isaac Greenwood, who was, incidentally, the first native American scientist in an American university to pursue postgraduate education abroad. Greenwood, born in America, was graduated from Harvard College in 1721. He went to England in 1723 and returned three years later. Toward the end of 1726, he announced a series of public lectures to be given in Boston. These lectures were delivered in the first months of 1727 and dealt largely with aspects of Newtonian mechanics. They appear to have been the first public scientific lectures given in the Colonies. The following year he became the inaugural Hollis Professor of Mathematics and Natural Philosophy at Harvard.

During the first four years of his tenure, Greenwood published scientific papers in the *Philosophical Transactions* of the Royal Society of London and communicated with various English scientists. He continued to give public scientific lectures in Boston during the 1730s, but his period of active scientific investigation had already ceased, a result no doubt of his having fallen victim to the "demon rum." In 1738 he was discharged from Harvard for continual drunkenness.

In the Minute Book of the Library Company of Philadelphia, we find the following extract under the date of 28 May 1740:

> At a Meeting of the Directors Present, Israel Pemberton Jnr., I. Stretch, W. Coleman, B. Franklin, P. Syng, E. Morgan, S. Rhoads, & Ph. Bond to consider a Request made by B. Franklin in Behalf of Issac Greenwood Professor of the Mathematics of Natural Philosophy, that they will lend him their Air-Pump, & allow him the Use of the Outer-Room adjoining to the Library, there to exhibit Mathematical & Philosophical Lectures & Experiments. The directors willing to encourage so useful a Design, agreed to grant the said Request for such a Time as will be sufficient for going thro' One Course of Experiments and Lectures.

The Minutes do not give any information as to whether or not the lectures were delivered. One may note that Greenwood's title has become confused in this extract; rather than Professor of Mathematics and Natural Philosophy, he has become Professor of the Mathematics of Natural Philosophy. The writer of the Minutes was probably thinking of Newton's great treatise, the *Principia,* "Mathematical Principles of Natural Philosophy," rather than of the title of the Hollis Professorship.

Franklin not only sponsored Greenwood's application, but acted as Greenwood's agent, just as he would do a few years later for Spencer, and collected the subscription fees at the post office. We know this from an advertisement in the *Pennsylvania Gazette,* no. 599, 5 June 1740, which incidentally seems to solve the problem of whether or not Greenwood actually gave his course of lectures:

> The Gentlemen who have subscribed to the *Encouragement* of a *Course of Philosophical Lectures and Experiments,* to be performed by Mr. Greenwood, are desired to meet in the *Chamber* adjoining to the *Library* at the *State-House,* on *Tuesday* next about 9 a Clock in the Morning, when it is proposed the *Course* should begin, and to be continued afterwards, at such Times as the Gentlemen who are willing to attend a *Course* at other Times, in the same Place, are desired to leave their *Names* at the *Post-Office* in *Philadelphia,* where the *Conditions* thereof may be seen, and *Subscriptions* taken in.

After leaving Philadelphia, Greenwood went to Charleston, where he engaged in private tutoring until he died in 1745.[38]

During the year 1744, Spencer was not the only person to give demonstrations in Philadelphia. The *Pennsylvania Gazette,* no. 811, 28 June 1744, carried an advertisement of an exhibition of "curiosities" which read as follows:

> *Near the upper End of Second-street, two Doors above the Sign of the Ship, is to be* SEEN, *in* John Baker's *House, after Six a Clock any Evening,* Among other CURIOSITIES, EIGHT BELLS ringing truly, both round Ringing and Changes, much in Imitation of Ringing in England, with two Young Men and a Lady walking, and the Lady turning Head over Heels, like a Mountebank, and the Clock drawing a Curtain. Also there is to be seen a Tower and Steeple, much like to that in Bristol in England, with 8 Bells and 8 Men ringing round Ringing and Changes, likewise an Image that blows a Bellows, strikes Fire, lights a Candle and fires a Gun.
>
> The Price of Men, 4d. Women 3d. Children 2d.
>
> The like never before in these Parts.

The exhibition of this mechanical contrivance may not be considered in the category of science, but it is worth giving in full since it places the next item in its true setting.

In the *Gazette,* no. 813, 12 July 1744, a small advertisement announced an elaborate set of spectacular demonstrations. It appeared again, somewhat enlarged, in the issue following, no. 814, 19 July; this was reprinted in no. 815 (misprinted as 814), 26 July. The first three issues for August (no. 816, 2 Aug.; no. 817, 9 Aug.; no. 818, 16 Aug.)

Juſt arrived from LONDON,

For the Entertainment of the CURIOUS and Others,

And is now to be SEEN, by Six or more, in a large commodious Room, at the Houſe of Mr. *Vidal*, in Second-Street;

The Solar or Camera Obſcura MICROSCOPE,

Invented by the Ingenious Dr. LIBERKHUN.

IT is the moſt Entertaining of any Microscope whatſoever, and magnifies Objects to a moſt ſurpriſing Degree. The Animalculæ in ſeveral Sorts of Fluids, with many other living and dead Objects, too tedious to mention, will be ſhewn moſt incredibly magnified, at the ſame Time diſtinct; alſo the Circulation of the Blood in a Frog's Foot, a Fiſh's Tail, alſo in a Flea, and Louſe, where you diſcover the Pulſe of the Heart, the moving of the Bowels, the Veins and Arteries, and many ſmall Inſects, that one Thouſand of them will not exceed the Bigneſs of a Grain of Sand, with their Young in them; Eels in Paſte, which have given a general Satisfaction to all that ever ſaw them. This Curioſity was never ſhewn by any Perſon that ever travelled. Price *Eighteen Pence.*

Each Seed includes a Plant, that Plant, again
Has other Seeds, which other Plants contain:
Theſe other Plants have all their Seeds, and thoſe,
More Plants again, ſucceſſively incloſe.
Thus, every ſingle Berry that we find
Has, really, in itſelf, whole Foreſts of its kind.
Empire and Wealth one Acorn may diſpenſe
By Fleets to ſail a Thouſand Ages hence.

Each Myrtle Seed includes a Thouſand Groves,
Where future Bards may warble forth their Loves:
So Adam's Loins contain'd his large Poſterity,
All People that have been, and all that e'er ſhall be.
Amazing Thought! what Mortal can conceive
Such wond'rous Smallneſs!----yet, we muſt believe
What Reaſon tells: for Reaſon's piercing Eye
Diſcerns thoſe Truths our Senſes can't deſcry.

Note, *The* Microscope *may be ſeen at Gentlemens Houſes, giving half an Hour's Notice, the Sun permitting only from Ten in the Morning to Four in the Afternoon, in my Room.*

THE unparallel'd MUSICAL CLOCK, made by that great Maſter of Machinery David Lockwood: This great Curioſity performs by Springs only; it is a Machine incomparable in its Kind; it excels all others in the Beauty of its Structure; is moſt Entertaining in its Muſick, and plays the choiceſt Airs from the moſt celebrated Operas, with the greateſt Nicety and Exactneſs: It performs with beautiful Graces, ingeniouſly and variouſly intermix'd, the French Horn Pieces, perform'd upon the Organ, German, and Common Flute, Flageolet, *&c.* as Sonata's, Concerto's, Marches, Minuets, Jiggs, and Scot Airs, compos'd by Corelli, Alberoni, Mr. Handel, and other great and eminent Maſters of Muſick. Price *Eighteen Pence.*

This beautiful Curioſity has been ſhewn twice before the KING, in his Royal Palace at St. James's, where His Majeſty was pleas'd to make an Obſervation on the Excellence of its Beauty, and declar'd, He thought it the Wonder of this Age. It is allowed by all who have ſeen it, to be more worthy to adorn a King's Palace than of being expos'd for a common Sight.

N. B. This ſurpriſing Piece of Machinery has given ſuch general Satisfaction to the Lovers of Art and Ingenuity, that the Nobility are continually commanding it to their Seats to ſatiſfy their Curioſity; and is to be SOLD by the Owner Edmund Rising.

The Inſide of this Machine may be view'd by Gentlemen and Ladies, and is to be ſeen from Eight in the Morning till Eight at Night.

For the Evening Diverſion,

THE Clock and Camera Obſcura, with the Battle of Dettingen, and ſeveral Italian Landſkips, repreſenting Armies, both Horſe and Foot, going through their Exerciſe at the Word of Command: Likewiſe Views of Ships fighting at Sea, with the Fiſh playing above Water, and Variety of Country Dances by Figures, ſix or eight Foot high, perform'd in a beautiful Manner by the Camera Obſcura. This Curioſity is eſteem'd one of the beſt Pieces of Painting of the Kind that ever was brought from *Italy.*

Fore Seat *Two Shillings.* Second Ditto *One Shilling.* To begin at Seven a Clock.

Broadside announcement published by Benjamin Franklin in 1744. Library Company of Philadelphia.

contained a notice of this attraction which occupied a whole single column. The details of the demonstration were displayed at length in a broadside printed by Franklin (see the reproduction). The popularity of the attraction is attested to by the continuing advertisements, which give evidence concerning the amount of scientific and quasi-scientific entertainment in Philadelphia during 1744.

But of all the lecturers on scientific subjects in Philadelphia in the middle of the eighteenth century, the most famous and the most popular was Ebenezer Kinnersley, a native Philadelphian. Kinnersley (1711–1778) was a Baptist minister who taught English and oratory at the College of Philadelphia, which has since grown into the University of Pennsylvania. Kinnersley was a co-experimenter of Franklin's and himself made several important discoveries, such as the conducting qualities of various sorts of charcoal; he also invented an "electric air thermometer."[39]

Kinnersley became famous for his lectures on electricity. He prepared these under Franklin's direction and, in fact, Franklin said in his autobiography that he wrote the greater part of them. Kinnersley delivered his first public lecture in Philadelphia in 1751 and in the same year went to Boston, where he repeated his lectures to an enthusiastic audience. The Bostonians liked these lectures so much that Kinnersley stayed on for several months repeating them. He gave the lectures again in New York City and in Newport, R.I., and then gave them in Philadelphia, on and off, for many years. In 1753, he went to Antigua, to St. Johns, and there too he gave his lectures on electricity.

Kinnersley's name was thus known all along the North Atlantic seaboard. It is even possible that during the 1750s his fame as an electrician was greater in America than Franklin's even though Kinnersley lectured chiefly on Franklin's discoveries. The primary difference between his lectures and those of Greenwood and Spencer is that he lectured exclusively on electricity, whereas they introduced the subject of electricity merely as one of a number of subjects of scientific interest. And the reason for this lies chiefly in the facts that between the time of their lectures and his, Franklin had made enough startling discoveries with Kinnersley's aid to provide materials for two full lectures on the subject of electricity, and that in Europe there had been developed an enormous stock of phenomena to supply a public lecturer with startling demonstrations.[40]

5

Collinson's Gift and the New German Experiments

In 1770, in a letter to Peter Collinson's son Michael, Franklin referred to the gift of "a Glass tube."[1] Franklin recalled the date to have been 1745, and he said that Collinson had sent (1) "an account of the new German Experiments in Electricity," (2) the "Glass Tube" in question, and (3) "some Directions for using [the glass tube], so as to repeat those Experiments." There is no record in the Library Company Minutes to make precise the date of arrival of Collinson's gift, but it was before 28 March 1747, the date when Franklin wrote a letter to Collinson, thanking him for his "kind present of an electric tube, with directions for using it," and telling him that his gift had been used with great success in "making electrical experiments."

Franklin's letter of thanks to Collinson mentions the tube and directions for using it, but the later letter to Collinson's son says that Collinson sent additionally what could have been either a printed description of the recent electrical discoveries of the Germans or a summary by Collinson of these discoveries. If it were a publication, then it would most likely have been an article entitled "An Historical Account of the Wonderful Discoveries, &c. concerning Electricity," published in London in the *Gentleman's Magazine* and printed again in Boston in the *American Magazine and Historical Chronicle*.[2]

Even if this article was not what Collinson had sent to the Library Company, it was certainly known to Franklin at the time that he began his electrical research—in concert with three fellow experimenters: Thomas Hopkinson, Philip Syng, and Ebenezer Kinnersley. I first called attention to the significance of this article in *Franklin and Newton*.[3] It was John Heilbron, however, who—in the course of his extensive research on electricity in the eighteenth century—found that Franklin's vocabulary and concepts were influenced by this publication: it was from this article that Franklin picked up the unusual term "electrise," which he used along with "electrify"; further, this was the source from which Franklin got the first inkling of the concept of an electrical "atmosphere" surrounding a charged body. It was also Heilbron who discovered that the article was a translation from a French text by the Swiss polymath Albrecht von Haller.[4]

Collinson's "directions" for using the glass tube are no longer extant. I believe that they took the form of a paper or letter containing handwritten instructions and perhaps a summary of (or a set of extracts from) the English version of von Haller's article. In fact, at about the same time, Collinson also sent such information to Cadwallader Colden in New York. We may, I believe, infer that Collinson's communication to the Philadelphia group would have been rather similar to his letter to Colden. The latter was based in large part on the account in the *Gentleman's Magazine,* supplemented by the actual experience of having seen some electrical experiments performed. The following pair of extracts shows the degree of similarity between Collinson's letter to Colden and von Haller's article:

> [Collinson's letter]
> As this may I think very justly be stiled an age of wonders, it may not perhaps be disagreeable to just hint them to you. The surpriseing phenomena of the polypus entertained the curious for a year or two past but now the vertuosi of Europe are taken up in electrical experiments . . .[5]

> [*Gentleman's Magazine*]
> What astonishing discoveries have been made within these four years! the polypus on one hand, as incredible as a prodigy, and the electric fire on the other, as surprising as a miracle!

When Franklin wrote to Peter Collinson's son about the gift of "an account of the new German experiments," a glass tube, and directions for using the tube to reproduce those experiments, his memory may have been faulty concerning detail. Recalling events of almost a quarter of a century earlier, he might very well have confused two simultaneous events, the receipt of a glass tube from Collinson with directions for using it and his own independent reading of the article in the *Gentleman's Magazine*. There would have been no need for Collinson to have sent this article to Philadelphia, as Heilbron has pointed out, since the Library Company then subscribed to this periodical and had in fact begun to do so at the specific suggestion of Collinson himself.[6] This latter fact indicates strongly that what Collinson would have sent was a handwritten description of the German experiments as he had seen them performed in England (possibly by William Watson), either as part of a letter or as an independent document.[7] Collinson's description was embellished by his reading about such experiments, notably in the *Gentleman's Magazine;* it reads in part as follows (letter to Colden):

> . . . and what can be more astonishing than that the base rub[b]ing of a glass tube should investigate [invest] a person with electric fire. He is not touched by the tube but the subtile effluvia that flies from it pervades every pore and renders him what wee call electrified, for then lett him touch spirits of wine & the spark . . . from his finger on the touch will sett the spirits in flame. This is a common experiment, but I have seen oyle of Sevil-oriangs [Seville oranges]—& camphire [camphor] sett on fire & gun powder mixt with oyl of lemmons will take fire—but what would you say to see fire come out of a piece of thick ice & sett the spirits in flame, or electrical fire drawn through water & performe the same [?]—these are some few of a great number of surprising things that are formed [performed?] by the electrical power, which you will find difficult to comprehend but are all facts.[8]

Whether or not Collinson sent to Philadelphia a copy of the article from the *Gentleman's Magazine* (either the whole issue or a cutting), what is of real importance is that Franklin's first report to Collinson on his electrical research shows that he had read this discussion on electrical experiment. For example, the article relates how "M. *Bose* . . . found that the kisses of a lady placed on the pitch, and electrified with the globe of glass, were as irksome as wounds by the pain which they caused." Franklin writes: "We encrease the force of the electrical kiss vastly, thus." In this letter Franklin does not refer directly to the article in the *Gentleman's Magazine,* although he does mention William Watson's *Sequel.*[9] Franklin stressed certain "particular phaenomena, which we looked upon to be new," although he was aware that "so many hands are daily employed in electrical experiments on your side the water," that someone or other "would probably hit on the same observations."

The reader will notice that the author of the article deals more with experiments than with theory, although there are references to "a whirl or vortex of electrical matter," to "the phlogistic or flammific force of electricity," and—above all—to an "electric atmosphere." He writes: "Electricity is a vast country, of which we know only some bordering provinces; it is yet unseasonable to give a map of it, and pretend to assign the laws by which it is governed." In his study of this article in relation to Franklin, Heilbron calls attention especially to this source of Franklin's own doctrine of "electric atmospheres" and observes that it was in this article that Franklin would have encountered the term "electrises," which he used at first (he later adopted "electrifies").[10] The "electric atmosphere," according to the *Gentleman's Magazine,* "when produced from large globes, extends itself four or five feet in circumference and agitates leaf-gold at that distance." Furthermore, said the author of that article, "I call it an atmosphere because it is really so, as appears by the smell, which naturalists have compared to that of oil of vitriol."

Heilbron has traced the source of this article, first published in French as "Histoire des nouvelles découvertes faites, depuis quelques années en Allemagne, sur l'électricité," in the *Bibliothèque Raisonnée des Ouvrages des Savans de l'Europe;* the author was Albrecht von Haller.[11] From this article, according to Heilbron, Franklin would have "learned of the fresh electrical games of the German professors: the ignition of warm spirits by sparks, the so-called electrical flare, first accomplished by C. F. Ludolff of Berlin; the electrical kiss and the catchy beatification invented by G. M. Bose of Wittenberg; the analogy between electricity and lightning, and the classification of sparks and glows, made respectively by J. H. Winkler and C. A. Hausen, both of Leipzig."[12]

It is no derogation of the originality of Franklin's researches to identify the experiments which he read about and then transformed into quite different experiments that yielded major new theoretical advances. In one of the experiments reported by Haller, and which Franklin would have encountered in the *Gentleman's Magazine,* an electrified man stands on pitch, surrounded by others who are on the ground: "Whoever shall approach his finger to the body of the person thus electrised will cause a spark to issue from the surface, accompanied with a crackling noise and a sudden pain of which both parties are but too sensible." Also, "If any other person not electrised puts his finger near one who is so, . . . there issues thence a fire, with a painful sensation, which both persons feel at the same time." Franklin's transformation was a radical one, consisting of placing two experimenters on wax insulating stands, while a third stands on the floor:

> 1. A person standing on wax, and rubbing the tube, and another person on wax drawing the fire, they will both of them, (provided they do not stand so as to touch one another) appear to be electrised, to a person standing on the floor; that is, he will perceive a spark on approaching each of them with his knuckle.
>
> 2. But if the persons on wax touch one another during the exciting of the tube, neither of them will appear to be electrised.
>
> 3. If they touch one another after exciting the tube, and drawing the fire as aforesaid, there will be a stronger spark between them than was between either of them and the person on the floor.
>
> 4. After such strong spark, neither of them discover any electricity.
>
> These appearances we attempt to account for thus: We suppose, as aforesaid, that electrical fire is a common element, of which every one of the three persons above-mentioned has his equal share, before any operation is

> begun with the tube. *A*, who stands on wax and rubs the tube, collects the electrical fire from himself into the glass; and his communication with the common stock being cut off by the wax, his body is not again immediately supply'd. *B*, (who stands on wax likewise) passing his knuckle along near the tube, receives the fire which was collected by the glass from *A*; and his communication with the common stock being likewise cut off, he retains the additional quantity received.
>
> —To *C*, standing on the floor, both appear to be electrised: for he having only the middle quantity of electrical fire, receives a spark upon approaching *B*, who has an over quantity; but gives one to *A*, who has an under quantity. If *A* and *B* approach to touch each other, the spark is stronger, because the difference between them is greater: After such touch there is no spark between either of them and *C*, because the electrical fire in all is reduced to the original equality. If they touch while electrising, the equality is never destroy'd, the fire only circulating. Hence have arisen some new terms among us: we say *B*, (and bodies like circumstanced) is electrised *positively; A, negatively*. Or rather, *B* is electrised *plus; A, minus*. And we daily in our experiments electrise bodies *plus* or *minus*, as we think proper.
>
> —To electrise *plus* or *minus*, no more needs to be known than this, that the parts of the tube or sphere that are rubbed, do, in the instant of the friction, attract the electrical fire, and therefore take it from the thing rubbing; the same parts immediately, as the friction upon them ceases, are disposed to give the fire they have received, to any body that has less.[13]

We may agree with Heilbron that "Franklin's ingenious idea to insulate the solicitor of sparks B as well as the electrified man A transformed Haller's little experiment into the prototypical demonstration of the existence of contrary electricities." This example provides a striking illustration of the doctrine that the most original, and even revolutionary, innovations in science are apt to be brilliant intellectual "transformations" of existing ideas, concepts, laws, theories, or experiments.[14]

6

The Kite, the Sentry Box, and the Lightning Rod

More than two hundred years have elapsed since the first experiments indicated that the lightning discharge is an electrical phenomenon. Yet today books and articles—on American history, on Franklin, on physics, and on the history of science—still exhibit so much uncertainty and even confusion about this subject that it is fitting to attempt to bring together the relevant printed and manuscript documents and to ascertain, insofar as possible, the sequence of events in Franklin's research on lightning in 1752.

The one fact about Benjamin Franklin's scientific career that is known to every reader of American history is that he once flew a kite during a thunderstorm; yet I am sure that most Americans would find extreme difficulty in answering the questions of precisely why he flew the kite and exactly what he learned in the process. Depicted in a famous Currier and Ives print, the familiar picture of Franklin raising his kite has become dear to generation after generation, while his magnificent contributions—both theoretical and experimental—to the budding science of electricity, which gained him in his lifetime the adulatory title of the Newton of the subject, have been ignored or forgotten.[1] Hence, it is not particularly surprising that the significance of the kite experiment for the development and acceptance of Franklin's ideas about electricity, or even lightning, has gradually become lost. Even those physicists who can readily understand what the experiment proved are apt to add such a statement as: "Perhaps the most wonderful part of it was that Franklin was not killed at once."[2]

The assertion has even been made that the kite experiment was never performed at all: "You have all heard the story of his kite-experiment, in which he [Franklin] got electric sparks between a kite-string and a key while the kite was flying in or near a thunder cloud. I regret to have to inform you that in the opinion of local historians this is just a myth, one of those legends which spring up from unknown sources to adorn the story of a great man."[3]

Rather than a myth or legend springing from unknown sources, the account of the kite experiment derives directly from Franklin himself. Even though there may be uncertainty about the actual day on which the kite was first flown, I do not believe that the performance of the experiment as such is subject to legitimate doubt.

In the following pages, I will present the evidence that the kite experiment was performed, not only by Franklin but by others. It may come as a surprise, however, to learn that this experiment had been independently conceived by a French experimenter, Romas. Furthermore, the electrical kite was the second test instrument that Franklin had devised to investigate the electrification of clouds; the earlier one was the insulated rod in the sentry-box experiment, which had been brought to a successful issue by European investigators even before Franklin raised his kite and which would have rendered the kite experiment unnecessary, had the news reached Franklin in time. I believe that the evidence fully supports Franklin's statement that the kite experiment was performed before he had heard of the success in Europe of his sentry-box experiment. We may conclude that the kite experiment was performed by Franklin in June 1752, and that the first grounded lightning rods were introduced to the world in Philadelphia in 1752 by Benjamin Franklin—probably in June, but possibly in July.

The Kite Experiment

The letter in which Franklin described his kite experiment was written from Philadelphia in October 1752. It was addressed to Peter Collinson, who had earlier provided Franklin with some simple apparatus for performing electrical experiments; many of Franklin's earliest contributions were addressed to Collinson.[4] A copy of the original letter, at present in the archives of the Royal Society of London, reads as follows:

> As Frequent mention is made in the Publick papers from Europe, of the Success of the *Philadelphia-Experiment* for drawing the Electrick Fire from Clouds by means of Pointed Rods of Iron erected on high Buildings &c., it may be agreeable to the Curious to be informed, that the same Experiment has succeeded in Philadelphia Tho' made in a Different & more Easie manner, which any one may try as follows:
>
> Make a small Cross of Two light Strips of Cedar, the arms so long as to reach to the four Corners of a Large Thin Silk Handkerchief, when extended; Tie the corners of the handkerchief to the extremities of the Cross; So you have the Body of a Kite, which being properly accomodated with a

> Tail, Loop, & String, will rise in the Air like those made of paper; but this being of Silk is fitter to bear the Wett & Wind of a Thunder Gust without Tearing.
>
> To the Top of the upright Stick of the Cross is to be fixed a very Sharp pointed Wire, riseing a foot or more above the Wood.
>
> To the end of the Twine, next the hand, is to be tied a Silk Ribon; and where the Twine & Silk joyn, a Key may be fasten'd.
>
> The Kite is to be raised, when a Thunder Gust appears to be comeing on (which is very frequent in this Country) & The Person who holds the String, must stand within a Door, or Window, or under some cover, so that the Silk Ribon may not be Wet; & care must be taken, that the Twine does not touch the Frame of the Door or Window.
>
> As soone as any of the Thunder Clouds come over the Kite, the pointed wire will draw the Electric Fire from them, & the Kite, with all the Twine, will be Electrified and the loose filaments of the Twine will stand out every way, and be attracted by an approaching finger.
>
> When the Rain has Wett the Kite & Twine, so that it can conduct the Electric Fire freely, you will find it stream out plentifully from the Key on the approach of your Knuckle.
>
> At this key the Phial may be Charged, and from Electric Fire thus obtained, Spirits may be kindled, and all the Other Electrical Experiments be performed, which are usually done by the help of a rubbed Glass Globe or Tube, & thereby the Sameness of the Electric Matter with that of Lightning compleatly demonstrated.
>
> I was pleased to hear of the Success of My experiments in France, & that they there begin to Erect points on their buildings. We had before placed them upon our Academy & State House Spires.[5]

This letter was read at the Royal Society on 21 December 1752 and was published in the Society's *Philosophical Transactions* for 1751 and 1752. It was also printed in the *Gentleman's Magazine* and the *London Magazine* and was included in the second part of Franklin's book on electricity. In America it was published in Franklin's *Pennsylvania Gazette,* and reprinted in the Boston *Gazette* and in other publications.[6]

A further and corroborative source of information is provided by Joseph Priestley's *History and Present State of Electricity.* This book has a special value, as Jernegan has shown, since Franklin was in close contact with Priestley and undertook to supply Priestley with the books he needed.[7] Furthermore, Priestley wrote that he had "kept up a constant correspondence with Dr. Franklin, and the rest of my philosophical friends in London; and my letters circulated among them all, as also every part of my history as it was transcribed."[8] Priestley's description appears, therefore, to have been based on information provided by Franklin himself and, since Franklin read the manuscript, must have had the seal of his approval. This account is more detailed than Franklin's letter to Collinson; for this reason Carl Van Doren preferred Priestley's description to Franklin's, and printed it in his edition of Benjamin Franklin's autobiographical writings, noting that it "contains precise details about the kite experiment which Priestley could have had only from Franklin, to whom Priestley refers as 'the best authority.' "[9] Priestley's account follows:

> As every circumstance relating to so capital a discovery (the greatest, perhaps, since the time of Sir Isaac Newton) cannot but give pleasure to all my readers, I shall endeavour to gratify them with the communication of a few particulars which I have from the best authority.
>
> The Doctor, having published his method of verifying his hypothesis concerning the sameness of electricity with the matter of lightning, was waiting for the erection of a spire in Philadelphia to carry his views into execution, not imagining that a pointed rod of a moderate height could answer the purpose, when it occurred to him that by means of a common kite he could have better access to the regions of thunder than by any spire whatever. Preparing, therefore, a large silk handkerchief and two cross-stocks of a proper length on which to extend it, he took the opportunity of the first approaching thunderstorm to take a walk in the fields, in which there was a shed convenient for his purpose. But, dreading the ridicule which too commonly attends unsuccessful attempts in science, he communicated his intended experiment to nobody but his son who assisted him in raising the kite.
>
> The kite being raised, a considerable time elapsed before there was any appearance of its being electrified. One very promising cloud had passed over it without any effect, when, at length, just as he was beginning to despair of his contrivance, he observed some loose threads of the hempen string to stand erect and to avoid one another, just as if they had been suspended on a common conductor. Struck with this promising appearance, he immediately presented his knuckle to the key, and (let the reader judge of the exquisite pleasure he must have felt at that moment) the discovery was complete. He perceived a very evident electric spark. Others succeeded, even before the string was wet, so as to put the matter past all dispute, and when the rain had wet the string he collected electric fire very copiously. This happened in June 1752, a month after the electricians in France had verified the same theory, but before he heard of anything they had done.[10]

On the basis of Priestley's report, we have further reason to believe that Benjamin Franklin performed the experiment as he described it in his letter to Collinson, and that the date was June 1752.

Stripped to its barest essentials, the kite experiment employed a long insulated conductor, terminating in a point at its uppermost end which was raised high above the ground; when a thundercloud passed overhead, Franklin momentarily grounded the insulated conductor by bringing his knuckle to the key, whereupon a spark passed between his knuckle and the key. The similarity between this experiment and the familiar experiments of the laboratory showed that thunderclouds are electrostatically charged; hence their discharge must be an electrical discharge, differing in scale, but not in kind, from those produced in the laboratory.[11]

The Sentry-box Experiment

The original experiment designed by Franklin to test the hypothesis that lightning is an electrical discharge between clouds, or between clouds and the earth, occurs in a paper entitled "Opinions and conjectures, concerning the properties and effects of the electrical matter, arising from experiments and observations, made at Philadelphia, 1749," and enclosed in a letter written by Franklin to Peter Collinson from Philadelphia, 29 July 1750.[12] It reads as follows:

> To determine the question, whether the clouds that contain lightning are electrified or not, I would propose an experiment to be tried where it may be done conveniently. On the top of some high tower or steeple, place a kind of sentry-box big enough to contain a man and an electrical stand. From the middle of the stand let an iron rod rise and pass bending out of the door, and then upright 20 or 30 feet, pointed very sharp at the end. If the electrical stand be kept clean and dry, a man standing on it when such clouds are passing low, might be electrified and afford sparks, the rod drawing fire to him from a cloud. If any danger to the man should be apprehended (though I think there would be none) let him stand on the floor of his box, and now and then bring near to the rod the loop of a wire that has one end fastened to the leads, he holding it by a wax handle; so the sparks, if the rod is electrified, will strike from the rod to the wire, and not affect him.[13]

It will be noted that if this experiment be stripped to its essentials, it is identical to the experiment of the kite; here, too, a pointed, insulated conductor is used to indicate whether thunderclouds overhead are electrically charged. In fact, the only real difference between the two experiments is that in the case of the kite the upper end of the pointed, insulated conductor is much higher in the air than the twenty or thirty feet above "some high tower or steeple" recommended by Franklin for the sentry-box experiment.

Franklin's description of the sentry-box experiment was published in his book on electricity, *Experiments and Observations on Electricity*, issued in London in 1751.[14] A French translation, made by Jean François

Franklin's sentry-box experiment. This drawing was made under Franklin's supervision, probably not by Franklin himself. From the "Bowdoin MS," American Academy of Arts and Sciences.

Dalibard (or d'Alibard) at the request of Buffon, was published in France in 1752.[15] This book created something of a sensation.

> The Philadelphian experiments . . . having been universally admired in France, the King desired to see them performed . . . His Majesty saw them with great satisfaction, and greatly applauded Messieurs Franklin and Collinson. These applauses of his Majesty having excited in Messieurs de Buffon, D'Alibard and de Lor, a desire of verifying the conjectures of Mr. Franklin, upon the analogy of thunder and electricity, they prepar'd themselves for making the experiment.[16]

The above description is taken from the report of the Abbé Mazéas, in a letter to Stephen Hales dated 20 May 1752 and read at the Royal Society on 28 May 1752. Mazéas then described Dalibard's experiments, in which the insulated rod had been erected in a sentry box in a garden at Marly-la-Ville; the "pointed bar of iron" was "40 feet high." During Dalibard's absence from Marly, the apparatus was entrusted to an "ancien dragon" named Coiffier. "On the 10 of May, 20 minutes past 2 afternoon, a stormy cloud having passed over the place where the bar stood, those, that were appointed to observe it [Coiffier and the village priest, Raulet], drew near, and attracted from it sparks of fire, perceiving the same kind of commotions as in the common Electrical experiments." Word was sent to Dalibard, who, on 13 May 1752, read a detailed account of the experiment, and the Franklinian principles of electricity it illustrated, to the members of the Académie Royale des Sciences in Paris.[17] On 18 May Delor repeated the experiment with similar success, using "a bar of iron 99 feet high, placed upon a cake of resin, two feet square, and 3 inches thick."[18] Before long, others had repeated the experiment—Le Monnier, Abbé Nollet, and others in France, Mylius and Ludolf in Germany, and, eventually, Canton, Wilson, and Bevis in England.[19]

If we accept Priestley's statement that the kite was flown in June 1752, then the Marly experiment of May 1752 antedated it by one month. This is perfectly consistent with Franklin's statement in the autobiography, when, after referring to the "capital [Marly] Experiment," he mentioned "the infinite Pleasure I receiv'd in the Success of a similar one I made soon after with a Kite in Philadelphia."[20] In 1768 Franklin wrote Dalibard a note in which he admitted freely that Dalibard was "the first of Mankind, that had the Courage to attempt drawing Lightning from the Clouds to be subjected to your experiments."[21] We must next investigate the question of whether, when Franklin flew his kite, he had already learned of the European experiments performed according to the Franklin specifications of 1750.

When Did Franklin Learn of the European Sentry-box Experiments?

The question of the date when Franklin received information about the successful issue of the sentry-box experiments performed in Europe has been studied by Abbott Lawrence Rotch,[22] Alexander McAdie,[23] Marcus W. Jernegan,[24] and Carl Van Doren.[25] Jernegan and Van Doren accepted Priestley's date of June 1752 for the kite, so that they endorsed the conclusion that Franklin had flown his kite before he had heard about the French experiments: Jernegan and Van Doren quoted Priestley's description of the kite experiment which I have printed above.[26] Priestley included in his history another discussion of this chronology: "Moreover, though Dr. Franklin's directions were first begun to be put in execution in France, he himself completed the demonstration of his own problem, before he heard of what had been done elsewhere: and he extended his experiments so far as actually to imitate almost all the known effects of lightning by electricity, and to perform every electrical experiment by lightning."[27]

This statement is even more explicit than the one quoted earlier. Even so, all writers have not been willing to accept Priestley's testimony. Thus Rotch believed that "the experiment was not performed until later in the summer" than June, [28] and McAdie concluded:

1. Franklin himself does not give a definite date when a kite was flown.
2. It seems doubtful that the kite was flown in June or early summer 1752.
3. If flown, the date was probably not far in advance of the end of September 1752.[29]

Rotch's and McAdie's conclusions would, according to Jernegan, "take from Franklin a part of the honor which he has had because of the belief that he independently made the discovery, by his own experiment, of the identity of lightning and electricity." His "kite experiment would rather be a continuation of those already performed; an attempt to extend them further." The expression "by his own experiment" is a little misleading; at the very least, it is ambiguous, since both experiments were Franklin's. Dalibard, in his report to the French Academy of Sciences on 13 May 1752, was explicit on this point: "In following the path that M. Franklin traced for us, I have obtained complete satisfaction."[30] Furthermore, it is difficult to know in what sense the kite experiment might have been conceived as "an attempt to extend" those experiments already performed, since the results it provided were identical to those of the sentry-box experiment.[31]

After reviewing the evidence, and especially on the basis of new information which he had uncovered, Jernegan concluded:

> Benjamin Franklin proposed the identity of lightning and electricity but by his own admission, a French scientist, M. D'Alibard, was the first to prove his conjectures by "drawing lightning from the clouds to be subjected to your experiments." On the other hand, the evidence presented makes it more certain that Franklin did prove the identity of lightning and electricity, *independently*, and that his "electrical kite" was flown in June, 1752, before he had heard of the French experiments of May 10 and 18. Secondly, while Franklin was the first to propose lightning rods, and to give definite direc-

> tions for erecting them, the evidence indicates that French scientists, acting on his suggestions, set up "sharp-pointed iron bars" on buildings and ships before May 26, 1752, before Franklin flew his kite, and before he himself had proved by experiment that they were a "preservative against thunder."

As I shall indicate below, there is still further evidence to support Jernegan's acceptance of Franklin's statement to Priestley that the kite experiment had been performed in June 1752 and, therefore, prior to his having received intelligence of the French experiments. On the other hand, I cannot agree that there is any good evidence that lightning rods had been erected in France earlier than those Franklin appears to have erected in Philadelphia in June 1752.[32]

Jernegan laid stress on a letter written by Franklin to Cadwallader Colden on 14 September 1752, in which he observed, "I see by Cave's *Magazine* for May that they have translated my electrical papers into French and printed them in Paris,"[33] since this issue of the *Gentleman's Magazine* contained a letter from France describing experiments "in pursuance of those by Mr. *Franklin* . . . to find whether the tonitruous and electrical matter be not analogous," and referring to "bars."[34] Jernegan therefore concluded that Franklin knew by 14 September that "French scientists had placed iron rods on buildings and that they were a preservative against thunder." The inference is that Franklin knew about the French experiments not long before writing his letter about the kite to Collinson in October; this point is also stressed by Van Doren.[35] A thorough examination of the *Pennsylvania Gazette* for the summer of 1752 has revealed that this information was in Franklin's hands almost a month earlier, that is, by the third or fourth week of August, since the issue for 27 August 1752 reprinted a letter concerning European lightning experiments with insulated rods.[36]

I am willing to accept Romas's computation to show that Franklin probably could not have received news of Dalibard's experiment at Marly-la-Ville in June 1752.[37] Dalibard's report was read in Paris on 12 May and Mazéas's letter to Stephen Hales was dated 20 May and was read at the Royal Society on 28 May. Since it would take a month or more for the news to reach Philadelphia from London, Franklin very likely would not have heard about this matter until early July—possibly the last days of June—at the very earliest, had either Collinson or Dalibard written him at once.

The account which Collinson eventually sent to Franklin was based on the Abbé Mazéas's letter, and reads:

> If any of thy Friends should take Notice that thy Head is held a little higher up than formerly, let them know; when the *Grand Monarch of France*

> strictly commands the *Abbé Mazéas* to write a Letter in the politest Terms to the Royal Society, to return the King's Thanks and Compliments in an express Manner to *Mr. Franklin* of *Pennsylvania,* for the useful Discoveries in Electricity, and Application of the pointed Rods to prevent the terrible Effects of Thunder-storms, I say, after all this, is not some Allowance to be made, if thy Crest is a little elevated? . . . I think, now I have stuck a Feather in thy Cap, I may be allowed to conclude in wishing thee long to wear it.

I have not been able to locate the original of this letter, but it plainly could not have been written more than a day or two before 28 May, the date when Mazéas's letter was read at the Royal Society, and more likely it was written after 28 May. The portion printed above was quoted (without its date) by Franklin in a letter to Jared Eliot dated 12 April 1753. This letter begins:

> The *Tatler* tells us of a Girl, who was observed to grow suddenly proud, and none cou'd guess the Reason, till it came to be known that she had got on a new Pair of Garters. Lest you should be puzzled to guess the Cause, when you observe any Thing of the kind in me, I think I will not hide my new Garters under my Petticoats, but take the Freedom to show them to you, in a Paragraph of our friend Collinson's Letter, viz.—But I ought to mortify, and not indulge, this Vanity; I will not transcribe the Paragraph, yet I cannot forbear.[38]

Surely this does not read like an introduction to an extract from a letter received by Franklin eight or ten months earlier. Rather, it gives the impression that Franklin had received Collinson's letter not too long before. That Collinson's letter had been in Franklin's hands when he wrote the kite letter in October seems, therefore, very unlikely, although we must keep in mind that we have no direct evidence one way or the other.

The earliest dated letter from Collinson to Franklin mentioning the French experiments was written on 7 July 1752, and begins as follows:

> I had the pleasure of my Dear friend's Letter of 21 March last with a guinea Inclosed but as I have Cash on hand I returned It by Moses Bartram. The Electrical Experiments have some thing very surprising in them, as all [thy letters] have. These our Friend Cave Intends to add to thy book as a Supplement and then the Errata may be added before they are printed. Wee Shall wait the Return of the Autumn or Spring Ships. It's likely our Friend Kinnersley may add some others, under thy Direction.
>
> By the Publick papers thou will See how thou has Sett the French to work.[39]

The last sentence above must refer to the accounts of the French experiments that were published in the *Gentleman's Magazine* or *London*

Magazine. In a letter of 27 September 1752, Collinson told Franklin briefly that "all Europe is in agitation on Verifying Electrical Experiments on points—all commend the Thought of the Inventor—more I dare not say least I offend Chast Ears."[40]

Franklin must have received some word of the Marly experiments from Dalibard himself, but I have been unable to find a copy of this communication. In the "avertissement" to the second edition of his translation of Franklin's book, Dalibard wrote:

> As soon as the first edition of this translation was completed, I sent a copy of it to M. Franklin, which put me into direct correspondence with him. I made known to him in time, the success of my experiment on thunder. I sent him the memoir which I had given to the Royal Academy of Sciences on 13 May 1752, such as it is in the second volume of the present work; he had been charmed by it and sent me, with his reply, his first supplement, the experiments of which I similarly verified. The second did not reach me until a long time afterwards.[41]

The earliest surviving letter I have found from Franklin to Dalibard is dated 29 June 1755.[42] The full text of this letter reveals that Franklin and Dalibard were never good correspondents. It begins:

> For a long time I have owed you a reply to your last letter, dated 20 June 1754. I received it last January while I was in Boston in New England,[43] & since that time I have been so busy with my travels to different places & with public affairs that I am extremely in arrears with my correspondents.
>
> I sent you last year a manuscript which contains some new experiments & some observations on thunder; I do not know whether you have received it, but it has since been printed in London, & I imagine that our good friend Mr. Collinson will have sent you a copy.
>
> I thank you for the kindness you have had in sending me the four volumes of natural history of M. de Buffon, the maps, &c.
>
> You desire my opinion of *Père Beccaria's Italian* book . . .[44]

In the fourth and fifth English editions of Franklin's book, this was printed as an "extract of a letter concerning electricity, from Mr. B. Franklin, to Mons. D'Alibard, at Paris, inclosed in a letter to Mr. Peter Collinson. F.R.S."[45] In a similar case, Collinson wrote Franklin on 20 July 1753 that "your Letter is forwarded to Mons. Dalibard by a safe Conveyance as soon as I received it & a Supplement with It—one I sent before to Mr. Buffon."[46]

But even though there is no way of determining the dates of the earliest Dalibard-Franklin correspondence, we may be certain that Franklin had not learned of the Marly experiments from him by mid-September 1752, and very likely even by October when he wrote the kite letter to Collinson. For, on 14 September 1752, when Franklin wrote to Colden that he saw in the May 1752 issue of the *Gentleman's Magazine* that a French translation of his book had appeared, he expressed the "hope [that] our Friend Collinson will procure and send me a Copy of the Translation."[47] We do not know exactly when Franklin received a copy, but sometime in the fall of 1753 he had received Nollet's volume of letters attacking his and had sent it to Colden. On 25 October 1753, he wrote Colden: "I send you herewith Nollet's book. M. Dalibard writes me, that he is just about to publish an answer to it, which, perhaps, may save me the trouble."[48] Colden's son David read Nollet's book and sent Franklin some comments on it which he included in the second supplement to his own book; in a letter to Colden of 1 January 1754, Franklin acknowledged the receipt of David's remarks, which had been enclosed in a letter from Colden "of the 3d past."[49] In this same letter Franklin mentioned his having received from London "the Supplemental Electrical Experiments" and noted that "Mr. Dalibard wrote to me that he was preparing an Answer" to Nollet.

The conclusion at which we arrive is that we have no way of telling when Collinson first wrote to Franklin about the success of the Marly experiment of Dalibard and those of Delor and Buffon, although it does not seem likely that he had mentioned this topic prior to his letter of 7 July 1752. Franklin wrote Colden on 14 September that he had just read that there was a French translation of his book; had he previously known from Dalibard about the Marly experiments, he would have certainly been told by Dalibard that he was the translator of his book. Hence, I believe that there is no reason whatever to doubt Franklin's statement, as reported by Priestley, that he had not known of the French experiments in June 1752, when he claimed to have flown his kite.

The Date of Franklin's First Lightning Rod

The version of Franklin's letter to Collinson describing the kite as printed in the *Philosophical Transactions* differs from that which appeared in his book on electricity and which has been used by the editors of collected and selected writings of Franklin, such as Sparks, Bigelow, and Smyth.[50] Whereas the date given in the *Philosophical Transactions* is 1 October 1752, the date given in Franklin's book and used by his editors is 19 October 1752. The latter is the date of the issue of the *Pennsylvania Gazette* containing the letter. Even more important than the date is the fact, first noticed by Hellmann, that the final paragraph, although printed in the *Philosophical Transactions,* was omitted from the version in Franklin's book.[51] It is also absent from the version in the *Pennsylvania*

Gazette and the various editions of his writings.[52] This paragraph states; "I was pleased to hear of the success of my experiments in France, and that they there begin to erect points upon their buildings. We had before placed them upon our Academy and state-house spires."

Jernegan assumed that Franklin was referring to a report given in a letter from Paris, written on 26 May 1752 and published in the *Gentleman's Magazine* for May 1752. This letter was printed immediately following a long extract from Franklin's book, dealing with his lightning hypothesis, which was described as: "A new Hypothesis for explaining the Phenomena of Thunder, Lightning, and Rain. Being an Extract from B. Franklin's *Experiments and Observations on Electricity.* Printed for E. Cave, and lately translated into French at Paris." Following the extract from Franklin's book, we find:

> *The above Hypothesis is in part confirmed by some Experiments lately made in France, as appears by the following Letter from Paris, dated May 26 N.S.*
>
> From several electrical experiments performed by our best naturalists, in pursuance of those by Mr. *Franklin* in *Philadelphia,* to find whether the tonitruous and electrical matter be not analogous, it appears, that to fix on the highest parts of buildings or ships sharp-pointed iron bars of ten or twelve feet, and gilt to prevent rust, with a wire hanging down on the outside to the ground, or about one of the ship's shrouds, is a preservative against thunder . . . The Sieur *Dalibard* having placed, in a garden at *Marly,* an iron bar on an electrical [i.e., insulated] body at the height of forty feet, was informed that on the tenth of *May,* about 20 minutes after two, a tempest passing over that spot, the parish priest and other persons drew from the bar such sparks and agitations as are seen in the common electrical performances. On the 18th the Sieur *de Lor* having fixed a bar at the height of 99 feet, on a cake of rosin, two feet square and three inches thick, drew coruscations from it during half an hour betwixt four and five, whilst the cloud was over it; these scintillations were perfectly like those emitted by his gun-barrel, when the globe is rubbed only with the brush, the same fire, the same crackling; whilst the rain, mixed with a little hail, fell from the cloud without any lightening or thunder, tho' it appeared to be the progress of a tempest which had happened elsewhere. Both these experiments have been reported to the Royal Academy of Sciences, and both evince that thunder clouds may be deprived of their fire, by iron bars fashioned and fixed as above.[53]

We know that Franklin had read this account before writing to Collinson in October since, as we have already seen, he referred to this issue of the *Gentleman's Magazine* in a letter to Colden dated 14 September 1752. We may note that the anonymous author of this Paris letter states that the experiments had been devised to discover whether "the tonitru-

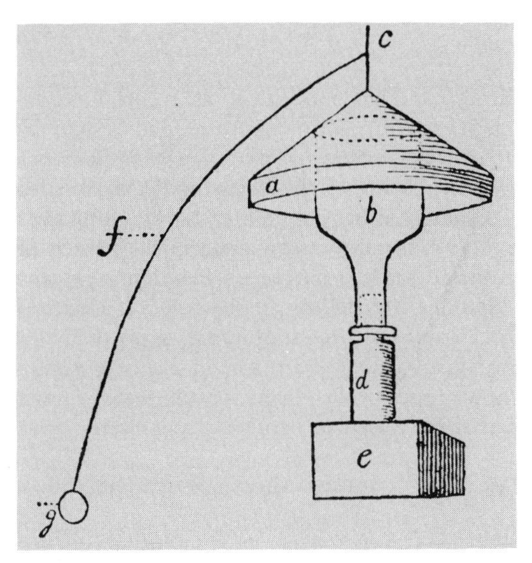

The *Gentleman's Magazine* for 1752 contained extensive discussions of Franklin's theory that lightning is an electrical phenomenon and various experiments demonstrating the electrification of clouds. One letter described the simple rooftop apparatus shown here, constructed of household objects and insulated with a glass bottle, which conducted electricity from clouds through a wire to the investigator. Harvard College Library.

ous and electrical matter be not analogous," and that the conclusion warranted by the demonstration of the "analogy" was that sharply pointed iron bars—affixed to the highest parts of buildings and ships and grounded—are a "preservative against thunder."

I have found evidence that Franklin knew about such experiments even earlier than 14 September 1752. The *London Magazine,* during this period, was reporting the news about electricity just as the *Gentleman's Magazine* was. The issue of the *London Magazine* for May 1752 has an almost identical letter from Paris, describing the work of Dalibard and Delor, which Franklin had before him when writing to Collinson on 14 September. The *London Magazine* for June 1752 carried a supplementary note on the French lightning experiments which reads as follows:

> *To what was said of Lightning and Electricity in our last . . . we shall add the following, which is also from Paris, June 12.*
>
> Tho' many very able and experienced naturalists have many years ago asserted, that lightning and the power of electricity were one and the same thing; which notion was grounded on the resemblance there was between their respective phenomena; yet resting satisfied with the conjecture only,

they never pointed out any ways or means for the demonstration of the fact. Mr. Francklyn, however, of Philadelphia in America, carried this critical point much further, and has pointed out the means for making the experiment; in which particular point he has succeeded beyond expectation. Mr. Lemonier, in particular, one of his most Christian majesty's physicians in ordinary, who is a member of the Academy Royal of Sciences, made the experiment accordingly at St. Germain en Laye, during the tempest which happened on the 7th instant; and planted in the garden of the Hotel de Noailles, an iron rod for that purpose.[54] He plainly perceived, that at the first flash of lightning that fell on it, the rod was electrified in the same manner, and had visibly the same appearance, as it would have had in case it had been electrified according to art. Abundance of persons of indisputable credit were eye-witnesses of the effects it produced; from whence it is now demonstrable, that the effects of lightening and electricity are the same.

The July 1752 issue contained three communications on lightning:

Further Remarks and Experiments in relation to Lightning and Electricity . . .

Paris, June 30. Upon the steeple of the church of Plauzat, in Auvergne, is a cross of iron, not painted or gilt. The extremities of this cross form sorts of fleurs-de-lis with sharp points. Whenever there happens any great storm, accompanied with thick clouds and flashes of lightening, a luminous body is perceived upon every one of the extremities of this cross. According to an immemorial tradition, there very rarely happens to be any thunder at Plauzat, or in the neighbourhood, when this phenomenon appears. As soon as it is seen, people are certain that the storm is no more to be feared. The luminous bodies are of different colours like the rainbow, and the figure is conical. Sometimes they continue an hour and an half, if it rains ever so plentifully.

Brussels, July 3. The Sieur Torre having caused a pointed iron rod to be erected upon the top of his house, on the 23d ult. at night, tho' there was but a slight appearance of a storm, shining sparks were drawn from that rod; on the 26th at night a dark cloud covered the sky, and a heavy rain, mixed with hail, fell when people were surprized to feel and to see, that a finger held at the distance of two inches from the rod, excited very strong sparks. These phaenomena greatly increased upon a clap of thunder being heard.

Paris, July 7. M. le Noine [Le Monnier], the king's physician, has made a new experiment in electricity, at St. Germaine en Laye, which confirms the analogy of the effect it has to that of thunder; the weather being very cloudy, he caused a cake of rosin to be brought to the place, upon which he mounted, and without any other instrument he extended his hand above his head, as a thicker cloud than ordinary passed over him, and one of those who were with him having touched him to make him remark something, he instantly received a most violent shock, of which fact he has made report to the Royal Academy of Sciences.

The same issue contained, in the section entitled "Foreign Affairs," further information on Dalibard's experiments:

> M. Dalibard, who frequently exhibits electrical experiments, got a bar of iron, or rather several joined together, to the length of 50 or 60 feet, erected at a village 7 or 8 miles from Paris, on the road to Compeigne: it was suspended by silken cords, and rested on glass bottles; so that supposing it could be electrified, it would not part with its virtue. One day a cloud passed over and discharged a clap of thunder, at which time M. Dalibard could draw sparks of fire from the bar, even at the distance of several inches. The flashes and sparks produced the pricking sensations as those from the conductor in the usual experiments. The diverging lucid stream was seen to issue from the pointed end of the bar; and every thing concurred to prove indisputably, that the bar was strongly electrified by the cloud. A gentleman who assisted at the experiment, upon slightly touching the rod unawares, received a violent stroke on his arm, and his clothes smelt all over of sulphur. The whole academy was entirely satisfied with the account, which clearly proved, that the matter of thunder and electricity is one and the same thing; and that it was practicable to extract thunder from a cloud, and direct it which way we please.

Franklin knew of these reports in the *London Magazine*. In the issue of the *Pennsylvania Gazette* for 27 August 1752, he published an "Extract of a Letter from Paris," taken from the May issue of the *London Magazine,* of which the text is substantially the same as that in the May issue of the *Gentleman's Magazine* which I have reprinted above. On 28 September, not long before he wrote to Collinson about the lightning kite, he published another account of lightning experiments in the *Pennsylvania Gazette,* this one being similar in content to the note printed above (from the July issue of the *London Magazine*). The text printed by Franklin is even closer to (though not a word-for-word reprint of) a note in the "historical chronicle" of the *Gentleman's Magazine* for July 1752. Franklin's note reads:

> Brussels, July 3. The Sieur Torre having caused a pointed Iron Rod to be erected upon the Top of his House, with Design, in some Measure, to dissipate the Fire which is in the Air, during the Time of a Storm, has succeeded therein beyond his Expectation; On the 23d of last Month, at Night, though there was but a slight Appearance of a Storm, shining Sparks were drawn from that Rod, but weaker than those drawn from an electrify'd Bar; on the 26th at Night a dark Cloud covered the Sky, and a heavy Rain, mix'd with Hail, fell, when People were surprised to feel and to see, that a Finger held at the Distance of two Inches from the Rod, excited very strong Sparks. These Phaenomena greatly increased upon a Clap of thunder being heard; insomuch that the Sparks grew considerably longer, larger and brighter. There is Reason to believe by this Experiment, that the pointed

> Rod or Bar may be of great Use in diminishing the Quantity of Fire from whence Thunder is formed, and in preventing the fatal Effects of Lightning.[55]

Hence, when Franklin wrote to Collinson in October 1752 about how glad he was to learn that the "French" were beginning to erect "points" on their buildings, he must have had in mind the item from Brussels that he had published in the *Gazette* sometime earlier. This note states that Monsieur Torre's rod had been erected early enough to enable him to perform lightning experiments on 23 June 1752.

Even more relevant is a note that Franklin later printed in the *Pennsylvania Gazette* (9 November 1752): "Paris, August 5. Several Persons of Quality have ordered Iron Rods to be fixed on the Tops of their Houses, to preserve them from the Thunder." As we shall see in the next section, it is not entirely clear that Torre's rod was designed *primarily* for protective purposes, whereas those referred to in the above note dated 5 August 1752 were to be erected solely for the purpose of preserving houses "from the Thunder." I cannot help feeling that this news item was in Franklin's hands when he wrote the letter about the kite to Collinson in October 1752. Franklin wrote about his pleasure in hearing of the success of his experiments in *France*—and I presume that he knew that Brussels was not in France—"and that they there begin to erect points on their buildings," a plain reference to more than one lightning rod erected on a building in France. Two months would have been sufficient time for news to have reached Philadelphia from Europe. Hence, I believe that the concluding sentence of Franklin's letter about the kite may legitimately be interpreted to mean that he had erected rods on two public buildings in Philadelphia earlier than the beginning of August 1752, that is, in June or July 1752.

This conclusion is consistent with the report by Priestley that the kite experiment was performed in June 1752; I assume that the lightning rods would not have been erected prior to experimental verification of the electrical nature of lightning, and there seems no indication of any other verification as early as June 1752 save the kite experiment.

Franklin's earliest suggestion of a lightning rod was included in a communication entitled "Opinions and Conjectures, concerning . . . Electrical Matter, Arising from Experiments and Observations Made at Philadelphia, 1749," and enclosed in a letter written by Franklin to Collinson dated 29 July 1750.[56] This is the same communication in which the sentry-box experiment was proposed. After describing a series of experiments in which charged insulated conductors had been discharged by a near-by pointed conductor, he wrote:

> . . . may not the knowledge of this power of points be of use to mankind, in preserving houses, churches, ships, &c. from the stroke of lightning, by directing us to fix on the highest parts of those edifices, upright rods of iron made sharp as a needle, and gilt to prevent rusting, and from the foot of those rods a wire down the outside of the building into the ground, or down round one of the shrouds of a ship, and down her side till it reaches the water? Would not these pointed rods probably draw the electrical fire silently out of a cloud before it came nigh enough to strike, and thereby secure us from that most sudden and terrible mischief?
>
> To determine the question, whether the clouds that contain lightning are electrified or not, I would propose an experiment . . .

Then followed the proposal to erect a sentry box on the top of a high tower or steeple.

It should be noted that the only question in Franklin's mind that demanded experimental proof was whether "the clouds that contain lightning are electrified." Once proofs were at hand that such is the case, then there would be no doubt whatever in his mind that lightning rods would work and would deprive the clouds of their electrical fire before they could discharge it in a bolt of lightning. Sufficient laboratory data had been accumulated to support his view; if the clouds were electrified and if, then, the lightning discharge was merely a bigger spark discharge than that obtained in the laboratory (not in any way different in kind), there was certainly no reason to suppose that the change in scale would affect the action of pointed conductors in discharging charged bodies—whether small metal laboratory objects or gigantic clouds.

To do the protective job assigned to them, lightning rods had merely to be fixed on "the highest parts" of "houses, churches, ships, &c." By contrast, the test rod in the sentry-box experiment was thought by Franklin to require "some high tower or steeple." In all likelihood, the greater elevation envisioned for the test rod was to ensure a large effect, since Franklin knew that a pointed conductor will "draw off" the charge from a charged insulated conductor with greater "ease" at near distances than from afar, and he obviously would have wanted the results of the sentry-box experiment to be on a sufficiently large scale to be convincing. According to Priestley, Franklin "was waiting for the erection of a spire in Philadelphia to carry his views into execution; not imagining that a pointed rod, of a moderate height, could answer the purpose." It has been assumed that the spire in question was that on Christ Church.

If we accept Priestley's word that Franklin flew his kite in June, and that this experiment successfully indicated the electrification of thunderclouds, then Franklin had no further reason to delay the introduction of lightning rods to protect buildings in Philadelphia.

What Kind of Lightning Rod Did Franklin Erect?

Many points of confusion exist with regard to early lightning rods. First, the distinction between grounded and ungrounded rods is not always made clear. Second, the action of the rods to prevent a stroke is not always kept distinct from their action in successfully conducting a stroke into the ground. Third, a considerable ambiguity exists about how proof may be obtained that the lightning rod is a "preservative against thunder."

In Franklin's communication of 1750, proposing the sentry-box experiment, he described two types of lightning rod.[57] One was grounded and its purpose was to "draw the electrical fire silently out of a cloud before it came nigh enough to strike, and thereby secure us from that most sudden and terrible mischief." We may note that such *protective* rods, as recommended by Franklin, were always to be grounded. The second type of lightning rod, to be used in the sentry-box experiment, was ungrounded or insulated. The reason why this *test,* or *experimental,* rod was not grounded is, of course, that its function was to become charged when a cloud passed overhead. In terms of Franklin's theory, a grounded rod would draw off the electrical fluid from clouds and transmit this fluid into the ground, an indefinite reservoir, until the charge on the cloud (its excess electrical fluid) was all removed; in a very short time, the cloud would be discharged and an experimental test would be difficult. Actually, as we shall see below, a grounded rod will become charged by induction when an electrified cloud is overhead, and such charge will be bound there so long as the cloud is above the rod; hence we know that the experiment can succeed as well with a grounded rod as an insulated one. Franklin's idea was to use an insulated rod, which would draw off some but not all of the electrical fluid of the cloud, which would remain on the rod somewhat longer since it would be dissipated not immediately into the ground, but only by the slower point discharge whereby electrified bodies "throw off" their excess electrical fluid if they are pointed.

Franklin was aware of possible hazard to a man standing near an ungrounded rod during a storm, as in the proposed sentry-box experiment, even though he stood on an insulating stand. He suggested, therefore, that the man might draw sparks from the charged insulated rod by means of a grounded wire with an insulating wax handle, so that "the sparks, if the rod is electrified, will strike from the rod to the wire, and not affect him." Even so, he declared his faith that, as to possible danger, "I think there would be none," and he did not take similar protective steps in performing the kite experiment. Richmann's death a few years later, while performing a variation of Franklin's sentry-box experiment, indicated that the danger in such experiments was greater than Franklin had envisaged; although we may note that when Richmann was electrocuted during a lightning storm, he was standing on the floor and not on an "electrical" (or insulated) stand.

The lightning experiments performed in France, Belgium, Germany, and England in 1752 were made with insulated rods or test rods, which had the function (by design) of indicating the electrification of overhead clouds, not of affording protection. As I pointed out earlier, Franklin was certain that a grounded pointed rod would afford protection from the lightning—if, that is, the thunderclouds were to prove to be electrically charged. Just as an insulated pointed conductor would draw off *some* of the electrical fluid from charged clouds, so a grounded pointed conductor would draw off *all* the electrical fluid from such clouds and render them harmless, that is, incapable of occasioning an electrical discharge, or lightning stroke. We have seen this idea expressed in his communication of 1750 (describing the sentry-box experiment). It also was stated in another letter of the same year (probably a little earlier), in which Franklin suggested that the end of the rod might, for greater effectiveness, be "divided into a number of points": "the electrical fire would, I think, be drawn out of a cloud silently, before it could come near enough to strike."[58] In "Poor Richard's Almanack" for 1753, announced in the *Pennsylvania Gazette* for 19 October 1752, Franklin merely indicated that the lightning would be attracted by the grounded rod and so preserve houses and ships from damage.[59]

The process whereby an insulated iron rod becomes charged is, according to our present state of knowledge, somewhat different from what Franklin envisioned. He assumed that if a cloud, say positively charged or with an excess of electrical fluid, passed directly overhead, then the insulated rod would draw off some of the electrical fluid from the cloud and become itself positively charged. We hold that the rod, in the presence of an overhead positively charged cloud, will exhibit positive electrification at the lower end and negative electrification at its upper end in a process known as "electrostatic induction"; incidentally, the first clear notions of electrostatic induction are due to Franklin, even though he did not apply them to this case.[60] If the electric potential is sufficiently great, some of the charge on the upper end of the rod will leak off, forming a glow discharge (which Franklin likened correctly to St. Elmo's fire or the "sailors' *corpusantes*"), and will move upward, being carried by ions (charged air molecules), and neutralize some of the charge on the bottom of the cloud. In discussing this phenomenon, B. F. J. Schonland, one of the foremost investigators of the lightning discharge, notes: "The process goes on until the rod has acquired a considerable excess of charge *of the same sign as that on the base of the cloud.* Franklin's experiment does not draw electricity from the cloud but has the same effect as if it had done so."[61] Hence, we see how a positively charged cloud (or a cloud whose lower portion is positively charged) causes an insulated rod to discharge negative electricity from its upper end, so as to be left positively charged and appear as if it had drawn off some electrical fluid from the positively charged cloud. In the same way, if a negatively charged cloud (or a cloud whose lower portion is negatively charged) passes over the rod, the upper end of the rod becomes positively charged, the ions streaming upward are positively charged, and the rod is left with a residual negative charge, just as if the cloud had drawn electrical fluid from the rod. Hence the sign of the charge on an insulated rod provides (as Franklin believed) a reliable index to the sign of the charge on clouds passing overhead, or at least the sign of the charge on the lower part of the cloud.[62]

In the case of the rod being grounded, we have a different situation. The "repelled negative 'induced' charge is no longer on the rod (having been repelled to a great distance) and the rod will discharge positive electricity so long as the cloud is near enough."[63] But, we must ask, does this phenomenon occur on a sufficiently large scale for the rod to be able to disarm the clouds of their charge and prevent a stroke? Schonland remarks on this score:

> As stated originally by Franklin, it [i.e., the lightning rod] depended for its success or failure upon the degree to which the upward discharge of electricity between pointed rod and cloud could render harmless the charge on the cloud. The lightning-rod, if it was to work in this manner, could only do so if the point discharge from it did actually neutralize the charge on the cloud to an appreciable extent. In the laboratory experiments which Franklin had made with his electrical machine, earth-connected metal points certainly neutralized electrified bodies placed near to them. But it would seem unlikely that the same thing could happen on the much larger scale of Nature when a puny point, a few tens of feet high, faced a thunder-cloud a mile or more above it.
>
> In actual fact it does not happen. A single point, or for that matter a multitude of points such as the tops of trees in a forest or the poles and chimneys of a town, has little effect upon the charge on the thunderstorm above it. None the less the lightning-rod, as is abundantly proven, has a very real virtue, because it "attracts" lightning to it . . . and can lead a flash to ground without damage to the building to which it is attached.[64]

Franklin was quick to learn that a lightning rod can also protect a ship or a building by attracting the lightning and safely conducting the charge into the ground. Many of those who objected to the lightning rods did not appreciate that Franklin advocated these two modes of action of rods: to prevent a stroke, and to conduct a stroke into the ground.[65] On 29 June 1755 he wrote to Dalibard that he had been "but partly understood in that matter":

> I have mentioned it in several of my letters, and except once, always in the *alternative, viz.* that pointed rods erected on buildings, and communicating with the moist earth, would either *prevent* a stroke, *or,* if not prevented, would *conduct* it, so as that the building should suffer no damage. Yet whenever my opinion is examined in *Europe,* nothing is considered but the probability of those rods *preventing* a stroke of explosion, which is only a *part* of the use I proposed for them; and the other part, their conducting a stroke, which they may happen not to prevent, seems to be totally forgotten, though of equal importance and advantage.[66]

We must now ask what kind of lightning rod Franklin had reference to when he wrote to Collinson in October 1752 that "points" had been erected on buildings in Philadelphia in June 1752. Were they ungrounded test rods as devised for the sentry-box experiment or were they grounded rods for protective purposes? In this letter to Collinson, Franklin wrote: "I was pleased to hear of the success of my experiments in France, *and* that they there begin to erect points upon their buildings." I have italicized the word *and,* since it emphasizes the sense of Franklin's letter as I read it, which seems to indicate that Franklin was pleased to learn that points were being erected upon buildings in addition to his pleasure at the success of his experiments. The points referred to by Franklin as having been earlier erected in Philadelphia were, I believe, designed to protect the Academy and the State House from lightning.[67] Franklin, I am sure, would not have exposed these two public buildings to any possibility of danger and he knew by the summer of 1752 that an ungrounded rod entailed some hazard; at the least, it did not provide the best possible protection. In his communication of 1750 to Collinson, and again in "Poor Richard" in the autumn of 1752, he insisted in the plainest terms possible that protective lightning rods be grounded.[68]

Although Jernegan believed that the reference to "points" on the Academy and State House implied that they were not for protection, but were rather test rods, he did not raise the question of whether they were grounded or insulated.[69] He argued that the appearance of the sentence about "points" erected in Philadelphia in a letter "concerned with the identification of lightning and electricity" implies that the "natural interpretation of these words would be that he placed 'points' on the 'Academy and state-house spires' for the purpose of experiment, and not for protection." But we have no indication of any experiment ever performed with these "points." Nor did Franklin ever imply elsewhere that he had used an insulated rod for lightning experiments before hearing of such experiments from Europe. Priestley insisted that Franklin had flown the kite before receiving the news of the European experiments, but neither he nor Franklin ever stated that he had performed the experiment with the insulated rod. Jernegan cited Priestley's statement:

Portrait of Franklin showing the lightning detector attached to the rod he erected in his house in September 1752. This portrait, reproduced from an engraving, was painted by Mason Chamberlin in 1762, when Franklin was fifty-six years of age. Burndy Library.

> The Doctor, after having published his method of verifying his hypothesis concerning the sameness of electricity with the matter of lightning, was waiting for the erection of a spire in Philadelphia to carry his views into execution, not imagining that a pointed rod of a moderate height could answer the same purpose; when it occurred to him that by means of a common kite he could have a readier and better access to the regions of thunder than by any spire whatever.

Jernegan concluded: "This would indicate that in June, the date given for the kite experiment, 'points' had not been erected on high buildings at that date even for experimental purposes." But, as Franklin was to learn from Europe, the experiment does not require that rods be placed as high as he had thought. If he had actually erected experimental rods, he would have performed experiments on them and he never claimed to have done so. Furthermore, he would not then have needed to erect yet another experimental rod, as he did on his house in September. Finally, I cannot see why he would have erected two experimental rods—one would have been ample. I would, therefore, agree that no experimental rods had been erected in Philadelphia in June 1752, but conclude that those on the Academy and State House must, therefore, have been protective or grounded rods.

In September 1752, Franklin erected a lightning rod on his own house in Philadelphia. It took an interesting form and it provided the means for making an important discovery. This rod was contrived "to draw the lightning down into my house, in order to make some experiments on it, with two bells to give notice when the rod should be electrify'd: a contrivance obvious to every electrician," as Franklin wrote to Collinson in September 1753.[70] This instrument[71] was described by Franklin in greater detail in his paper "Experiments, observations, and facts, tending to support the opinion of the utility of long, pointed rods, for securing buildings from damage by strokes of lightning, read at the committee appointed to consider the erecting of conductors to secure the magazines at Purfleet, August 27th, 1772":

> In Philadelphia I had such a rod fixed to the top of my chimney, and extending about nine feet above it. From the foot of this rod, a wire (the thickness of a goosequill) came through a covered glass tube in the roof, and down through the well of the staircase; the lower end connected with the iron spear of a pump. On the staircase opposite to my chamber door, the wire was divided; the ends separated about six inches, a little bell on each end; and between the bells a little brass ball, suspended by a silk thread, to play between and strike the bells when clouds passed with electricity in them.[72]

Franklin's description of this rod as a device "to draw the lightning down into my house, in order to make some experiments on it" has, I believe, misled the modern historian. Jernegan observed, "It is clear, however, that this was not a rod set up for protection but for experimental purposes." Yet Franklin's description of the action of the rod indicates that, in addition to its use in experiment, it could provide adequate protection:

> After having frequently drawn sparks and charged bottles from the bell of the upper wire, I was one night awaked by loud cracks on the staircase. Starting up and opening the door, I perceived that the brass ball, instead of vibrating as usual between the bells, was repelled and kept at a distance from both; while the fire passed, sometimes in very large, quick cracks from bell to bell, and sometimes in a continued, dense, white stream, seemingly as large as my finger, whereby the whole staircase was inlightened as with sunshine, so that one might see to pick up a pin.[73]

Since the separation of the bells was about six inches, any quantity of lightning sufficient to do damage to an unprotected house could arc across this air gap and so be carried from the upper member of the rod to the lower member and be successfully conducted into the ground without causing any damage. Franklin knew that his rod would afford protection to his house on the basis of his investigations of the path followed by lightning when buildings were struck.[74] Thus, in his letter to Collinson describing the lightning rod with the bells, he wrote:

> In every stroke of lightning, I am of opinion that the stream of the electric fluid, moving to restore the equilibrium between the cloud and the earth, does always previously find its passage, and mark out, as I may say, its own course, taking in its way all the conductors it can find, such as metals, damp walls, moist wood, &c. and will go considerably out a direct course, for the sake of the assistance of good conductors; and that, in this course, it is actually moving, though silently and imperceptibly, before the explosion, in and among the conductors; which explosion happens only when the conductors cannot discharge it as fast as they receive it, by reason of the being incomplete, dis-united, too small, or not of the best materials for conducting. Metalline rods, therefore, of sufficient thickness, and extending from the highest part of an edifice to the ground, being of the best materials and complete conductors, will, I think, secure the building from damage, either by restoring the equilibrium so fast as to prevent a stroke, or by conducting it in the substance of the rod as far as the rod goes, so that there shall be no explosion but what is above its point, between that and the clouds.

One of the best known instances of Franklin's tracing the path of lightning in order to show how the lightning will depart from a simple path "to pass as far as it can in metal" occurred when the Newbury

(Mass.) church was struck. This was described in a letter Franklin wrote to Dalibard on 29 June 1755. But Franklin was aware of this phenomenon at least as early as August 1752, when he published in the *Pennsylvania Gazette* for 6 August the following account:

> Last Friday, early in the Morning, the Lightning struck two Houses on Society Hill, and did them considerable Damage, but hurt no Person. It was very remarkable in both Houses, that the Lightning in its Passage from the Roof to the Ground, seem'd to go considerably out of a direct Course, for the sake of passing thro' Metal; such as Hinges, Sash Weights, Iron Rods, the Pendulum of a Clock, &c. and that where it had sufficient Metal to conduct it nothing was damag'd; but where it passed thro' Plaistering or Wood work, it rent and split them surprizingly.[75]

I believe that the lightning rods erected in Philadelphia in 1752 were the first grounded lightning rods to be erected anywhere in the world for the purpose of protecting buildings from the lightning discharge. Dalibard's rod at Marly-la-Ville was erected in the middle of a garden and was insulated. Delor's rod was erected on his house in Paris, but it too was insulated, as was the apparatus used by LeMonnier, Mazéas, LeRoy, Cassini de Thury, and the Abbé Nollet. In England, Canton erected an insulated instrument built for the occasion of a tin tube with needles attached to the top and Wilson performed the experiment with an iron curtain rod projecting out of the window. Mylius and Ludolf in Berlin likewise used insulated rods.

The letters in the *Gentleman's Magazine* and the *London Magazine* often referred to the fact that experiments with insulated rods had proved the efficacy of lightning rods, but such statements do not imply that the insulated test rods were afterward grounded in order to become effective protective rods, nor that rods for protection had been erected.[76] Jernegan quotes the letter from the *Gentleman's Magazine* for May 1752:

> From several electrical experiments performed by our best naturalists, in pursuance of those by Mr. *Franklin* in *Philadelphia,* to find whether the tonitruous and electrical matter be not analogous, it appears, that to fix on the highest parts of buildings or ships sharp-pointed iron bars of ten or twelve feet, and gilt to prevent rust, with a wire hanging down on the outside to the ground, or about one of the ship's shrouds, is a preservative against thunder.

He then concludes that Franklin, who had read this letter, "knew by September 14, 1752 that there was a report that French scientists had placed iron rods on buildings and that they were 'a preservative against thunder.' " But the "report" does not warrant this conclusion. Plainly, it seems to me, the report implies that since experiments (with ungrounded conductors) have proved that lightning clouds are electrified, or that the "tonitruous and electrical matter" are "analogous," then (as Franklin has pointed out, since electrified clouds must follow the same laws as electrified bodies in the laboratory) "it appears" that a pointed grounded conductor "is a preservative against thunder."

An understandable confusion arises, of course, from the very nature of the experiments with insulated rods. They appeared to draw some "tonitruous matter" from the clouds and thereby become electrified. But, in drawing off "tonitruous matter," did they not lessen the striking power of the cloud overhead and so offer some protection? In this light, we can understand the beginning of the report from Brussels, which Franklin published in the *Pennsylvania Gazette,* in which Monsieur Torre's rod was said to have been erected "with Design, in some Measure, to dissipate the Fire which is in the Air," but the remainder of the communication does not make it clear whether the rod was grounded or insulated. Yet this account is written in terms so similar to current reports on insulated rods, such as those printed above, that I cannot help feeling that his rod, too, may have been insulated.

The Paris report of 5 August 1752 stated merely that several persons of quality had ordered protective rods to be erected on their houses, and if Franklin's specifications were used, these rods would have been grounded. I have been unable to find when (or, for that matter, whether) these rods were actually erected.

The "divided" lightning rod with bells that Franklin erected on his house in September 1752 afforded adequate protection and, if my reading of the final sentence in the kite letter be correct, two public buildings in Philadelphia were protected by rods in June or July 1752.

Eight years later, when Franklin was in England, he wrote Kinnersley that, despite the importance of the lightning rod "on our side of the water . . . Here it is very little regarded; so little, that though it is now seven or eight years since it was made publick, I have not heard of a single house as yet attempted to be secured by it."[77] Two months earlier, on 24 January 1762, he had sent the philosopher David Hume a description of the method for constructing lightning rods, omitting "the philosophical reasons and experiments on which this practice is founded; for they are many, and would make a book. Besides they are already known to most of the learned throughout Europe." Although the "philosophical reasons and experiments" were known throughout Europe, the practice was evidently not. "In the American British colonies," continued Franklin, "many houses have been, since the year 1752, guarded by these principles."[78]

Franklin's Delay in Reporting the Kite Experiment (and Other Subjects of Doubt)

If, as I have attempted to show, there is no reason to doubt that Franklin had conceived and executed the kite experiment before hearing the news of the French performance of the sentry-box experiment, the reader may well inquire why so much space is necessary for the investigation of this question. Why should we not simply accept Franklin's and Priestley's word: especially since Franklin's statement in his letter to Collinson of October 1752 and Priestley's account form a consistent picture with all of the evidence cited above? The answer lies in the fact that the June date has been seriously questioned and some investigators have concluded that Franklin must have flown the kite much later, at a time when he had already heard of the successful conclusion of the sentry-box experiment in France.[79] Furthermore, the statement has been made more than once that Franklin did not perform the experiment at all, or that, if he did, he did not report his results accurately in the letter to Collinson, which is then supposed to be a plan of an experiment to be made rather than an account of one that has already been made. Even further, as I pointed out in the beginning of this essay, a statement has been made to the effect that the experiment is nothing but a myth.

The most ardent critic of the kite experiment was Alexander McAdie, who concluded his article on the date of Franklin's kite experiment with the following statement: "The whole tenor of the letter of October 1 (19) 1752, indicates not so much an experiment actually performed as one projected and the results anticipated. For actually the phenonema are quite different. Franklin does not say in the concluding paragraph that he actually charged a phial, etc. Only that it may be charged."[80]

McAdie gave a particular drubbing to Dr. Stuber, pointing out that his "account is in general terms; and what is rather surprising, explanatory and apologetic." Stuber's account is as follows:

> It was not until the summer of 1752, that he was enabled to complete his grand and unparalleled discovery by experiment. The plan which he had originally proposed, was, to erect on some high tower, or other elevated place, a sentry-box, from which should rise a pointed iron rod, insulated by being fixed in a cake of resin. Electrified clouds passing over this, would, he conceived, impart to it a portion of their electricity, which would be rendered evident to the senses by sparks being emitted, when a key, the knuckle or other conductor was presented to it. Philadelphia at this time afforded no opportunity of trying an experiment of this kind. While Franklin was waiting for the erection of a spire, it occurred to him that he might have more ready access to the region of clouds by means of a common kite. He pre-

> pared one by fastening two cross sticks to a silk handkerchief, which would not suffer so much from the rain as paper. To the upright stick was affixed an iron point. The string was, as usual, of hemp, except the lower end which was silk. Where the hempen string terminated, a key was fastened. With this apparatus, on the appearance of a thunder-gust approaching, he went out into the commons, accompanied by his son, to whom alone he communicated his intentions, well knowing the ridicule which, too generally for the interest of science, awaits unsuccessful experiments in philosophy. He placed himself under a shade, to avoid the rain—his kite was raised—a thunder-cloud passed over it—no sign of electricity appeared. He almost despaired of success, when, suddenly, he observed the loose fibres of his string to move towards an erect position. He now presented his knuckle to the key, and received a strong spark. How exquisite must his sensations have been at this moment. On this experiment depended the fate of his theory. If he succeeded, his name would rank high among those who had improved science; if he failed, he must inevitably be subjected to the derision of mankind, or, what is worse, their pity, as a well-meaning man, but a weak, silly projector. The anxiety with which he looked for the result of his experiment, may be easily conceived. Doubts and despair had begun to prevail, when the fact was ascertained in so clear a manner that even the most incredulous could no longer withold their assent. Repeated sparks were drawn from the key, a phial was charged, a shock given, and all the experiments made which are usually performed with electricity.[81]

I must agree with McAdie's acerb comment that if by the statement "he placed himself under a shade" Stuber implied that Franklin stood under a tree to escape rain, then Franklin would not have been heeding his "own previously published warning that it was very dangerous to stand under trees during a thunder-storm."[82] On the other hand I am not so sure that I agree with McAdie's stricture: "Why should one who had made an estimate of what we may call the killing power of lightning, wish to expose his own son to probable death or at any rate intense shock?" I am sure that Franklin did not believe that he was taking his own life and that of his son into his hands when he performed this experiment. We may note, in this regard, that in his description of the original experiment of the sentry-box, he had stated that he did not believe there was any danger of the experimenter being killed by the lightning discharge during the performance of the experiment. McAdie points out that Franklin would hardly have gone to the common if he had desired to fly his kite "where none could see and comment." But he was thinking of a New England common in the center of the town, whereas the Philadelphia common was on the outskirts.

McAdie directed his criticism at Stuber, we may note, rather than at Priestley. "Dr. Stuber knew Franklin intimately and it is said got the story of the kite from him," he declared. How reliable a witness was Stuber? An early nineteenth-century edition of Franklin's works referred to him as "one of the Doctor's [i.e., Franklin's] intimate friends."[83] Sparks stated cautiously that Stuber, "who resided in Philadelphia . . . seems to have written from minute and accurate information."[84] Parton stated, "We owe our knowledge of what occurred on the memorable afternoon [on which the kite was flown], to two persons who had heard Franklin tell the story, namely, Dr. Stuber of Philadelphia and the English Dr. Priestley."[85] (What warrant there may be for placing the experiment in the afternoon, I do not know.) George Simpson Eddy, who provided McAdie with much of his information concerning Stuber, stated: "I have an edition of the Life and Essays of Dr. Franklin published in the Republic of Letters, a journal which was published in New York in . . . 1834. This life begins in No. 2 of that journal, page 171. On page 180 begins the continuation of Franklin's life written by Stuber, who is described as 'one of the Doctor's intimate friends.' "[86] Eddy apparently was willing to accept this statement that Stuber was a close personal friend of Franklin and did not express any disapproval of Parton's statement that Franklin had told the story in person to Dr. Stuber.

I do not know what the evidence is that Dr. Stuber was an intimate friend of Franklin, nor even that he got the story at first hand from Franklin.[87] There is no correspondence extant between Stuber and Franklin; he is not mentioned in any Franklin letter in the Smyth, Bigelow, or Sparks editions of Franklin's writings, nor was he of sufficient importance to be mentioned in Carl Van Doren's biography. The only information contained in Stuber's account which is not to be found in Priestley's is open to serious question, as McAdie pointed out, and leads one to believe that Stuber merely added a few personal embroideries to the account which he had obtained from Priestley.

But even if we may not place too much confidence in Stuber, I see no reason to doubt the credibility of Franklin's and Priestley's testimony. McAdie could not believe that Franklin had flown his lightning kite in June because no account of the experiment appeared in the newspapers: it seemed to him "quite improbable that a man so astute as Franklin and so keenly aware of the importance of this particular experiment, would have failed to publish a note, however brief, and preliminary, in the *Gazette.*" Yet the evidence indicates that Franklin was not in the habit of publishing brief and preliminary notes about his scientific discoveries in the *Gazette;* in fact, the publication of the kite letter in the issue of October 19 is, so far as I have been able to tell by an examination of the files of the *Gazette* during the years in which Franklin made his experiments, a solitary exception. And even in this case we may note that a communication had gone off to Collinson (and through him to the Royal Society).

McAdie raised another doubt:

> What is perhaps still more significant, E. Kinnersley, who was the chief expositor of the newly-discovered electric fire, and who was in close correspondence with Franklin (Franklin borrowed his "brimstone globe" March 2, 1752, and used it in making experiments in the spring of 1752) gave several public lectures, in which there is no mention of the kite experiment. In the Pennsylvania Gazette of September 14, 1752, there is an account of Kinnersley's lecture at the State House. And again in the issue of September 21, September 28, and October 19.[88]

What reason can there have been for Franklin to give Kinnersley the opportunity of making the first announcement of the lightning kite in Philadelphia? Considering the importance of the experiment, would we not rather have expected Franklin to have reserved for himself the first public statement about the kite, as he apparently did?

Finally McAdie noted, "It would also seem that, once assured of the results, Franklin would have wasted no time in communicating with Peter Collinson to have the paper laid before the Royal Society." There are, I believe, two plausible reasons why Franklin might have delayed his report from June to October.

First, we must remember that Franklin was often slow (especially by then-current standards) in sending reports of his experiments to England. Thus the paper entitled "Further Experiments and Observations in Electricity" was enclosed in a letter to Peter Collinson dated Philadelphia, 29 April 1749, beginning: "Sir, I now send you some further experiments and observations in electricity made in Philadelphia 1748 . . ."; the famous paper entitled "Opinions and Conjectures" was not sent to Collinson until 29 July 1750; and in the letter written to Collinson in September 1753 (about the negative electrification of clouds) Franklin noted his dilatoriness at the very beginning when he wrote, "In my former paper on this subject, written first in 1747, enlarged and sent to England in 1749 . . ." The letter last mentioned, written in September 1753, described experiments which Franklin performed during the period from September 1752 to 6 June 1753.

A second possible reason for the delay, it seems to me, is that Franklin may very well have hoped for another opportunity to repeat the experiment before writing a full report for Collinson and the Royal Society. We know from a letter which Franklin wrote to John Perkins on 13 August 1752 that his affairs were pressing, that "business sometimes obliges one to postpone philosophical amusements."[89] The kite experiment was an almost incredible performance, and Franklin knew it. He might, therefore, well have deemed it necessary to have performed this experiment at least once more before publishing a formal account of it. This would certainly explain why Franklin did not write an account of the kite to Collinson immediately after hearing of the news about the French experiments, which, as I have shown above, occurred toward the end of August.

Yet another reason for the delay has been advanced by Carl Van Doren. According to Van Doren, might not Franklin "deliberately have kept his secret till October so that he might publish at the same time, or almost the same time, in his newspaper and in his almanac the two most important pieces of his year's news? That is what he did. On 19 October his first account of the *Electrical Kite* appeared in the *Gazette*. The same issue advertised as in the press the new *Poor Richard* for 1753, which contained Franklin's first positive statement of *How to Secure Houses, etc., from Lightning*."[90] The chief weakness in this argument, however, is that a description of lightning rods had been published in the first edition of Franklin's book on electricity, which had already been out for a year.

Very likely, the delay on Franklin's part was caused by a multitude of factors, and the most important may have been a fear of ridicule. Many a modern commentator has called this experiment foolhardy. Others still find it remarkable that Franklin and his son were not electrocuted. It certainly would have taken a great deal of courage for anyone to have said he had actually drawn down the lightning from the sky, or even that he had dared to fly a kite during a thunderstorm. Priestley points out, in the report which had its origins in Franklin's statements to him, that Franklin, "dreading the ridicule which too commonly attends unsuccessful experiments in science . . . communicated his intended experiment to nobody but his son, who assisted him in raising the kite." To be sure this is a reference to the possibility of ridicule in case of failure, yet the notion of ridicule was clearly present in Franklin's own mind—even when he told Priestley about the experiment years later.

There can be no question that it was difficult to take seriously a proposal to test the electrification of clouds by drawing down the lightning from heaven. The June 1752 issue of the *Gentleman's Magazine* contained "a letter from a gentleman at Paris to his friend at Toulon, concerning a very extraordinary experiment in electricity, dated May 14, 1752," which began:

> You must remember, Sir, how much we ridiculed Mr. *Franklin's* project for emptying clouds of their thunder, and that we could scarce conceive him to be any other than an imaginary Being. This now proves us to be but poor *virtuosi;* for yesterday I met a learned gentleman of the academy, who assured me that the experiment had been very lately tried with success. You

may suppose I could scarce think him serious; however, I found that a memoir read at one of their assemblies had made so extraordinary an impression upon him, that I began myself to abate of my incredulity.

This gentleman was probably not alone in his sentiments, and I am sure that there were many who not only ridiculed Franklin's sentry-box experiment when it was first proposed, but who also could scarcely believe it when they were told that such an experiment had been successful. The British members of the Royal Society of London, despite the approval by some of them of Franklin's lightning hypothesis, did not make the experiment that Franklin had proposed. We do not know exactly why, and the reason may well have been that the "connoisseurs" laughed at it.[91] The laughter is all the more remarkable when we consider the splendid reception his earlier papers had received, even his preliminary statements on the electrification of clouds.[92] One use made of Franklin's conception of the electrification of clouds, as Collinson had written to him in April 1750, was by those who wished to solve the phenomena of earthquakes.[93]

The news that man had been able to draw the lightning from the skies was certainly astounding. Could the reports be true? We must remember that the first news of the lightning experiment originated in France; how many agreed with Collinson's sentiments in a letter to Franklin of 20 July 1753: "Wee know the French Very Well, subject to Levity, Hights & Extreams"?[94] When Watson published in the *Philosophical Transactions* of the Royal Society an account of the experiments with the insulated rod made by John Canton, Benjamin Wilson, and John Bevis, he noted that the effects were "trifling" when "compared with those which we have received from Paris and Berlin, but they are the only ones, that the last summer here has produced." Nevertheless, "as they were made by persons worthy of credit, they tend to establish the authenticity of those transmitted from our correspondents."[95] Surely the last part of this sentence indicates that there was not a universal trust placed by Englishmen in the reports from France and Germany.

It seems to me, therefore, that an important reason why Franklin did not at once make public the results of the lightning experiment in June was the fear that no one would take him seriously; he did not want to compromise his reputation. After he had heard of the news from France, he was then willing to publish a brief account of the kite, since he now had independent confirmation of what he had proved by means of this experiment.

No one who reads the pages of the *Pennsylvania Gazette* for the period from June to October 1752 can help noticing the many references to lightning, chiefly accounts of storms and their destructive effects during the early summer and news of European experiments in the late summer and early fall. Quite obviously, the subject of lightning was on Franklin's mind, as it well might have been if he had already performed the kite experiment and had erected two lightning rods in Philadelphia. One such account, originating from Portsmouth, New Hampshire, 23 July 1752, and published in the *Gazette* for 6 August, was followed by a postscript of Franklin's, reading: "A plain Proof of the Electrick Nature of Lighting." This postscript is all the more interesting in that the notice preceding it did not actually indicate such a proof at all, and simply stated "The main Mast of a Schooner at the North-end was struck by the Lightning; and altho' the Mast was shiver'd to Pieces by it (and the other Mast ruined by the Shock) till it came to a Ring that encompassed it (which it melted a little) yet below that Ring there were no Effects of it—A plain Proof of the Electrick Nature of Lightning."

Does this not read as if Franklin, having proved "the Electrick Nature of Lightning" (by the kite experiment), could not help adding a conclusion which, if not warranted by the facts reported from New Hampshire, was uppermost in his mind? He certainly knew what a "proof" was.

If we accept Priestley's statement that the kite experiment was performed in June, then Franklin's letter to Collinson of October 1752 indicates that he did not hesitate long before erecting rods on at least two buildings in Philadelphia.

The kite experiment had proved to his own satisfaction that thunderclouds are electrified, or are charged bodies; hence, according to everything that he had learned about electricity, the lightning rods should work. On this basis, we can understand why he did not include, in the version of his letter on the kite which he published in the *Pennsylvania Gazette,* the final paragraph about the erection of "points"; I believe that this topic is fully understandable as explained by Van Doren: "The rods on the Academy and State House were already known to Philadelphia."[96]

It is well to reemphasize that once Franklin knew that clouds were electrified, he was certain that the lightning rods must work as he had predicted, since there was no reason to suppose that electrified clouds would behave differently with regard to pointed grounded conductors than ordinary electrified bodies in his laboratory. Jernegan and others indicate that Franklin had not in 1752 "proved by experiment that they [lightning rods] were a 'preservative against thunder.' " The only real proof to be had of the efficacy of lightning rods in preserving houses against lightning required that a bolt of lightning hit the rod and not destroy the house. The first such occasion arose eight years after the lightning rods had been first erected, in 1760, when the house of Mr.

West was struck and was saved from destruction by the lightning rod which had been erected on it.[97]

If we can understand why Franklin might have held back any public announcement of the kite experiment from June until October, we must still answer certain other objections raised by McAdie. One of them is that Franklin did not repeat the kite experiment. Once Franklin had heard of the French experiments, which provided all the information he desired, further experiments with a kite were unnecessary. In September 1752, as we saw previously, he erected a form of lightning rod in his own house which provided him with ample means of making experiments on electrified clouds. The statement in his autobiography makes clear that Franklin thought that his sentry-box experiment as performed by Dalibard was superior to the kite experiment; certainly it was easier to perform. Once Franklin had heard, late in August 1752, of the European experiments, he immediately (September 1752) constructed the dual-purpose rod (to serve as a protective instrument for his house and as an experimental tool), since he now knew that the test instrument did not have to be as high as he had originally supposed. With this instrument, he easily verified for himself the results obtained in Europe and those previously obtained by him with the kite. Having now more than once proved the electrification of thunderclouds, he wrote a note to Collinson about the kite.[98] All of Franklin's subsequent research on lightning was done with the rod, although (as we shall see in the next section) Kinnersley made further experiments with the kite.

Romas's Claim to Priority

Among those who performed experiments with lightning kites in the mid-eighteenth century, one of the most interesting is the Frenchman Jacques de Romas, *assesseur au présidial de Nérac,* since he claimed priority in the invention of this instrument.[99] In a book supposedly dealing with lightning conductors, and the means of protecting houses from thunder, he devoted a considerable amount of space to establishing the grounds for his independent discovery. He related that "the first experiment on the electricity of thunder [at Marly] was announced to the public by all the gazettes and other periodical works,"[100] and he then decided to repeat these experiments, not—as he says—because he doubted their veracity, but rather to see whether there were new phenomena to be explored, which might be important "for the utility of civil society, or the progress of physics." He made some experiments with an insulated rod or bar but, wishing to increase the effects, he "plunged himself into

One of the many experiments made by Romas with his electrical kite. Frontispiece to his *Mémoire sur les moyens de se garantir de la foudre dans les maisons* (1776). Burndy Library.

meditation." Finally after one half hour the idea of the kite (*le cerf-volant des enfants*) presented itself to his mind. In a letter which he wrote to the Académie de Bordeaux on 12 July 1752 he announced his plan to use as a means of exploring the electrification of clouds "un Jeu d'enfant."[101] However, August passed and the time of thunderstorms was over. He therefore waited until after the following winter and did not raise his kite until 14 May 1753, while a second experimenter watched the insulated rod erected on his house so that the two types of observation might be coordinated.

Romas insisted that "un Jeu d'enfant" referred unambiguously to a kite, and I see no reason to doubt his word. The paramount question in Romas's mind was whether Franklin had actually performed his kite experiment in June 1752, as Priestley said he had in his history of electricity, or whether Franklin had done it later. Plainly, to establish his own priority, Romas had first to show that Priestley's attribution of the month of June must be an error. But even assuming that Franklin's experiment had been made at the end of June, Romas wanted to prove that it would have been impossible for him to have received news of it earlier than the thirteenth of July; in other words, even if Franklin had thought of the kite earlier than he had, at least he wanted credit for independent invention. Romas asked: if Franklin had known him personally and had sent him a special message about the kite experiment at the end of June, could it have arrived in Bordeaux in as little as thirteen days? But he had never heard tell of Benjamin Franklin in June 1752 "and I do not have enough vanity to flatter myself that at this same time I had the honor to be known to him,"[102]—to say nothing of the possibility of a ship getting the message from Philadelphia to Bordeaux within thirteen days. The first news of the alleged kite experiment of Franklin ("la prétendue expérience du Cerf-volant de M. Franklin"[103]), according to Romas, arrived in the hands of his London correspondents only by January 1753[104] and did not arrive in France until the fifteenth of January when Watson wrote a letter to Nollet about it; how then "could I have been informed about it by 12 July 1752?"

Romas had heard of the Marly experiment by reading an account in the *Gazette de France* for 27 May 1752, a copy of which arrived at Nérac, where he was stationed, only in the first days of June.[105] He assumed that at least an additional month would have been required to get the news to Philadelphia. Hence, if—as he believed—Franklin flew his kite only after hearing of the experiments of Dalibard and Delor, he could not have flown his kite in June as Priestley had asserted. Romas claimed, furthermore, that Priestley himself had indicated that Franklin's kite experiment postdated his learning of the French sentry-box experiments. The quotation which follows was alleged by Romas to have been taken from the 1767 (first English) edition of Priestley's history, and it was repeated verbatim by Merget in a supposedly definitive article on Romas written for the Academy of Bordeaux in the nineteenth century.

> Mr. Franklin is the first person (we now know what to believe regarding this priority) who suspected the identity of lightning and electric fluid; he indicated beforehand the way to establish this identity when he proposed insulating in the open air, during a storm, a pointed rod that could be electrified by contact; the first electrical display effected by this device occurred in France under the eyes of Mr. de Lor and Mr. d'Alibard. *Mr. Franklin, animated by the success of these two gentlemen,* himself tested the success of his rod in Philadelphia, where he then was. This physicist, having also had lucky success, soon reflected that by means of a kite he could obtain surer and easier access to the region where lightning is produced. This method was found to be correct through the trial which he made of it in the month of June in that same year 1752 in the countryside of Philadelphia, where he thought it best to proceed with no other witness than his son, in order to avoid the derision of fools.[106]

It will be noted that the italicized phrase exactly contradicts the sense of Priestley's own words. This text as a whole does not come from Priestley's history at all, although several phrases (for instance, the end of the final sentence) do; I suspect that Romas was quoting from a French review of Priestley and had never seen the original at all.[107]

Romas tells us that on 19 October 1753 he addressed a letter to Benjamin Franklin,[108] along with two memoirs.[109] Franklin replied in a letter dated 29 July 1754, which Romas later printed in English and also in French translation. Although Franklin gave him "the hope that he would write again, I have never received any other letter."

On the score of Franklin's letter to him, Romas wrote: "That which is worthy of being remarked well in this letter is that M. Franklin makes no claim to the invention of the kite [experiment]. That was, however, the time when he should have done so: he must have perceived in my letter, and more clearly still in the first memoirs [that I sent him], that I claimed to be the originator of this instrument."[110] As a matter of fact, added Romas, he had also thought of making lightning experiments using an insulated bar (but one that ended in a ball rather than a point) in 1750, which was "more than a year before M. Franklin."[111]

Watson wrote a letter to the Abbé Nollet under the date of 15 January 1753, which was printed in French translation in a footnote to one of the two memoirs sent by Romas to Franklin, beginning: "M. Franklin has sent to the Royal Society, a fortnight ago, a very pretty electrical experiment for drawing electricity from the clouds."[112] There followed a description of the construction of the kite, precautions to be taken with it (including some not in the letter as printed in the *Philosophical Transactions*) and the experiments to be made with it. The note containing this letter then goes on to state: "It seems by this letter that M. Franklin has used the kite prior to M. de Romas; but judging by the same letter and by the memoir of the latter, one will see that the effects were much greater at Nérac than at Philadelphia. This difference comes, it would appear, from the fact that M. de Romas garnished the cord of his kite with a metal wire, as one will see by reading his memoir."

If Franklin had flown his kite in June 1752, as I have every reason to believe he did, then Romas did not conceive of the same experiment until a month later, and since he did not fly his kite until 14 May of the following year, he thereby lost priority to Franklin both in the invention of the experiment and in its performance. If Romas had actually conceived the experiment of erecting an insulated rod in 1750, it is a pity that he did not at once describe what he proposed to do and the means for doing it, since this would have given him priority in devising the first experiments to draw the lightning from the skies. Since he neither published the idea, nor made the experiment prior to Dalibard's Marly experiment of May 1752, we are hardly entitled to give him credit for it. On the other hand, Romas may well have thought of using an insulated rod to test the electrification of clouds independently. As Franklin noted after quoting to Lining the extract from his diary about how he "came first to think of proposing the experiment of drawing down the lightning, in order to ascertain its sameness with the electric fluid," the thought "was not so much 'an out-of-the-way one,' but . . . might have occurred to any electrician."[113] In any event, the possibility—even the probability—that lightning is an electrical phenomenon was not new in Franklin's day; what was original was an experiment to test this oft-expressed idea, and Franklin's was the first to be made public and to be carried into execution.

Although the note in Romas's memoir indicated that he had conceived the kite experiment independently of, but later than, Franklin, Romas—perhaps on the basis of Franklin's not asserting his own claim vigorously—concluded that Franklin might not have performed the experiment at all. At any rate, we know that in 1764 he asked the Academy of Sciences to adjudicate his claim to priority. Two commissioners, Nollet and Duhamel, reported: "Having regard to all these proofs, we believe that M. de Romas had not borrowed from any one the idea of applying the kite to electrical experiments, and that one must regard him as the first author of this invention, until M. Franklin or some other makes known by sufficient proofs that he had thought of it before him. (4 February 1764.)"[114]

A nineteenth-century partisan of Romas, Merget, noted:

> With his ordinary prudence, Franklin . . . remained with his mouth closed, as if he recognized on his part the justice of the judgment of the Academy; but this sly resignation did not prevent him, three years later, in 1767, from letting his friend Priestley speak of Romas in cavalier terms which we have transcribed above. One can allege, it is true, in his justification, that he was ignorant of the declaration of the commissioners of the Academy; this is very possible without being in any way probable. But that which is beyond doubt, in any case, is that he knew in their extent the claims of his competitors; as the latter, under the date of 19 October 1753, had sent him two memoirs where these claims were very clearly expressed, and where the experiment of the lightning kite, recounted in all its details, is presented as an original experiment.[115]

Franklin, it is true, never entered the lists in order to defend his own claims to the prior invention of the lightning kite. In scientific matters, his procedure was always that expressed by the lawyers' phrase *res ipsa loquitur*. In his autobiography, he related that he had not personally answered any of the attacks made on his ideas by the Abbé Nollet, having "concluded to let my Papers shift for themselves; believing it was better to spend what time I could spare from public Business in making new Experiments, than in Disputing about those already made."[116] At the height of the controversy in England as to whether lightning rods should end in balls or points, he wrote to Le Roy from London (30 March 1773) that "I have an extreme Aversion to Public Altercation on Philosophic Points, and have never yet disputed with any one, who thought fit to attack my Opinions."[117] A few months later, he wrote to Ingenhousz that he would not answer a pamphlet by Wilson "against Points . . . being averse to Disputes."[118] When Ingenhousz was embroiled in a dispute with Priestley over the problems of photosynthesis, Franklin wrote to him:

> I hope you will omit the polemic piece in your French edition and take no public notice of the improper behaviour of your friend; but go on with your excellent experiments, produce facts, improve science, and do good to mankind. Reputation will follow, and the little injustices of contemporary labourers will be forgotten; my example may encourage you, or else I should not mention it. You know, that when my papers were first published, the Abbé Nollet, then high in reputation, attacked them in a book of letters. An answer was expected from me, but I made none to that book, nor to any other. They are now all neglected, and the truth seems to be established. You can always employ your time better than in polemics.[119]

"Whatever some may think and say," he wrote to Ingenhousz, "it is worth while to do men good, for the self-satisfaction one has in the reflection."[120]

In this spirit, he undertook no dispute with Romas. And, in a later publication, he generously referred to Romas's experiments:

> M. de Romas saw still greater quantities of lightning brought down by the wire of his kite. He had "explosions from it, the noise of which greatly resembled that of thunder, and were heard (from without) into the heart of the city, notwithstanding the various noises there. The fire seen at the instant of the explosion had the shape of a spindle, eight inches long and five lines in diameter. Yet, from the time of the explosion to the end of the experiment, no lightning was seen above, nor any thunder heard. At another time the streams of fire issuing from it were observed to be an inch thick and ten feet long."[121]

Priestley devoted considerable space to Romas in his history and noted: "The greatest quantity of electricity that was ever brought from the clouds, by any apparatus prepared for that purpose, was by Mr. De Romas, assessor to the presideal of Nerac. This gentleman was the first who made use of a wire interwoven in the hempen cord of an electrical kite . . ."[122]

Franklin's friend and co-experimenter Kinnersley also performed kite experiments. They are described in a letter written to Franklin on 12 March 1761:

> Whether the electricity in the air, in clear, dry weather, be of the same density at the height of two or three hundred yards, as near the surface of the earth, may be satisfactorily determined by your old experiment of the kite. The twine should have throughout a very small wire in it, and the ends of the wire, where the several lengths are united, ought to be tied down with a waxed thread, to prevent their acting in the manner of points. I have tried the experiment twice, when the air was as dry as we ever have it, and so clear that not a cloud could be seen, and found the twine each time in a small degree electrized positively. The kite had three metalline points fixed to it; one on the top, and one on each side. That the twine was electrized, appeared by the separating of two small cork balls, suspended on the twine by fine flaxen threads, just above where the silk was tied to it, and sheltered from the wind. That the twine was electrized positively, was proved by applying to it the wire of a charged bottle, which caused the balls to separate further, without first coming nearer together. This experiment showed, that the electricity in the air, at those times, was denser above than below. But that cannot be always the case; for, you know, we have frequently found the thunder-clouds in the negative state, attracting electricity from the earth; which state, it is probable, they are always in when first formed, and till they have received a sufficient supply. How they come afterwards, towards the

latter end of the gust, to be in the positive state, which is sometimes the case, is a subject for further inquiry.[123]

Nor was Kinnersley the only American to repeat Franklin's experiment; another was John Lining. In a letter dated 14 January 1754, Lining described his kite in the following terms:

> The kite, which I used, was made in the common way; only, in place of paper, I covered it with a silk, called *alamode*. The line was a common small hempen one of three strands. A silk line, except it had been kept continually wet, would not conduct the electricity; and a wire, besides other inconveniences, would have been too heavy. I had not any instrument, whereby I could take the height of the kite; but, I believe, it was at least 250 feet high. It was flown in the day-time.[124]

Lining evidently used a key at the end of the kite string, just as Franklin had, and he charged a Leyden jar from it, repeating the experiments usually made with the electrical machine at its prime conductor.

Another mid-eighteenth-century physicist to experiment on atmospheric electicity with a kite was the Abbé Beccaria, a staunch Franklinist, who did much to promote the use of lightning rods in Italy and who appears to have been the first person to be successful at electrolyzing metallic compounds. In one of his writings on atmospheric electricity, he related:

> It was in the year 1756, that the frequent and continued use of kites, which other observers only used to make researches on the electricity of clouds, procured me a confirmation of what I had till then only conjectured, that is to say, that even during clear weather (except in the cases of a great dampness of the air, or of an impetuous wind) a mild weak electricity perpetually took place.
>
> Kites were most useful instruments to me, for such first experiments on the state of the atmosphere. They rise to a great height, to a region where the difference of the atmospheric electricity uses to be greater; they gather great quantities of this electricity, by means of the pack-thread which holds them, and they retain it the better as they are capable of being insulated . . . Now, a string made of the best silk, of a small diameter, and of great length insulates a kite extremely well; and it is an easy matter to keep it dry by warming it, or to change it, when it grows damp. Though I was at first ignorant of the contrivance of Sig. Romas, who interweaves the string which holds his kite with thin metallic wires, the same thought occurred to me the more naturally, as I was then exploring the accidents of the weaker electricity that takes place in serene weather.[125]

According to Priestley, Beccaria

> made use both of kites and pointed rods, and of a great variety of both at the same time, and in different places. Some of the strings of his kites had wires

> in them, and others had none. Some of them flew to a prodigious height, and others but low; and he had a great number of assistants, to note the nature, time, and degree of appearances, according as his views required.
>
> To keep his kites constantly insulated, and at the same time to give them more or less string, and for many other purposes, he had the string rolled upon a reel, which was supported by pillars of glass; and his conductor had a communication with the axis of the reel.[126]

Yet another to make experiments on atmospheric electricity with a kite was Peter Van Musschenbroek, one of the discoverers of the Leyden jar.[127] It should be noted, however, that apparently none of these experiments was made during a severe lightning storm, that in all cases the key and the kite string or wire were charged by electrostatic induction, and that the kite was used to advance knowledge of the electrification of the atmosphere even in serene weather. Thus the kite provides us with an example of the way in which a tool invented to solve a specific problem (the possible electrification of clouds) finds application in investigating a much larger class of phenomena.

Conclusion

Because electrical kites were flown by others, there is no reason to suppose that one might not have been flown by Franklin. Priestley's testimony, approved by Franklin, set the date of the experiment in June 1752, one month after the performance of Franklin's earlier (sentry-box) experiment. The implication of Franklin's letter of October 1752, describing the kite to Collinson, is that some sort of metal "points" had been erected in Philadelphia in June 1752, hence earlier than the rod (with the warning bells) erected by Franklin on his own house in Philadelphia in September 1752.

No information has as yet been uncovered, despite a considerable research by a number of different individuals, to make us reject the testimony of Franklin and Priestley. Although a number of perplexing questions have been raised, they may all be answered in a reasonable way, although some conjecture is required on occasion—for example, to explain the delay in publishing the experimental results. Further confidence in the accounts by Franklin and Priestley arises when we review the accumulated collateral material on lightning and lightning rods in the period from June to October 1752, all of which is consistent with the dates and other information they provided. In any event, it is difficult to think that Franklin, who was always exceedingly honest in reporting scientific information, would have grossly falsified the record.

I do not believe that all relevant information about this episode has been uncovered. Everyone who has ever investigated the question has been able to add something new. Perhaps the discovery of hitherto unknown Franklin correspondence, or of diaries and letters of his friends or fellow inhabitants of Philadelphia, will some day reveal a few more details about the kite itself and the Philadelphia lightning rods. But, as matters stand now, a little more than two hundred years after the time when Priestley asserted with Franklin's approval that the kite was flown, we may with confidence conclude that Franklin performed the lightning kite experiment in June 1752, and that soon after, in late June or July 1752, it was in Philadelphia that the first lightning rods ever to be erected were put into service.

7

Father Diviš and the First European Lightning Rod

with Robert Schofield

Questions of priority and independence of discovery often arise in the history of science and it usually proves difficult to resolve them completely. Occasionally a document may turn up in which one of the contestants has referred explicitly to the scientific work of one or more of the others, thereby enabling us to prove indebtedness. In the absence of such a document, however, the historian may find it impossible to establish positive proof of indebtedness, but can occasionally provide information as to the probability of independence. Such a situation exists in the case of the curious lighning rod which Father Procopius Diviš (also Diviss or Diwish) erected in Moravia on 15 June 1754 and which is mentioned, in passing, in a number of histories of physics and of electricity.

The two-hundred-fiftieth anniversary of the birth of Diviš was marked in 1948 by celebrations in Czechoslovakia and the publication, in Czech, of a sumptuous monograph by Karel Černý, containing the Czech text of Diviš's *Magia naturalis,* and a preface (in English, Czech, Russian, and French) by Karel Hujer.[1] Hujer subsequently produced an account of the facts concerning Diviš and his lightning rod, summarizing the extensive primary and secondary literature in the Czech languages. Hujer[2] presents the views held by a number of historians of physics, namely: the lightning rod erected by Father Procopius Diviš in 1754 was the first grounded or protective lightning conductor to be erected in Europe; Father Diviš conceived the idea of a grounded lightning conductor in complete independence of Franklin; and the form of the lightning rod erected by Diviš was unique.[3] We believe, however, that the evidence indicates that not all of these claims made for Diviš can be accepted without serious question.

We may note, at the outset, that in all probability neither Franklin nor Diviš would be very much impressed by our efforts to establish claims for one or the other. Franklin abhorred polemics and verbal controversies and always believed it better to spend what time he "could spare from public business in making new experiments than in disputing about those already made."[4] A quotation from Diviš cited by Hujer indicates that he had a similar viewpoint. Diviš seems to have reluctantly accepted the fact that he was not generally considered the first inventor of the lightning rod. He wrote that although his "meterorological machine" (or lightning rod) was of great significance as evidence for his own theory, he himself was nevertheless aware that Franklin, "on account of his universal reputation," was generally accorded priority in this invention.[5]

Yet, the evidence to be presented in the following pages may throw light, not only on the questions of independence and priority, but also on the speed of transmission of scientific information and the acceptance of the theory of the lightning rod on the Continent in the mid-eighteenth century; we may, therefore, be justified in ignoring the presumed wishes of the principals in this dispute and reviewing the whole issue.

That Diviš should be recognized as one of that group of experimenters who worked, often in the face of a considerable popular opposition, to establish the usefulness of lightning rods is, as Hujer shows, beyond question. This is not, however, a unique example of a church official supporting the use of lightning rods in opposition to an uninformed and conservative populace. For example, when Giuseppe Verratti encountered great local resistance to the lightning rod he had erected in Bologna, the Pope himself, Benedict XIV, wrote a letter advocating the use of lightning rods; we may note that in Bologna as in Přímětice, prejudice won the day.[6]

According to Hujer, Diviš's manuscripts indicate that he had "the plan of his lightning conductor all complete by the end of 1752 or the beginning of 1753." Hence we may inquire into the general state of knowledge with regard to lightning and lightning conductors at that time. Diviš, as presented by Hujer, appears to have been a competent experimental scientist. It seems unlikely, therefore, that an experimenter of his ability would have erected a lightning rod without first knowing, by experimental proof, that clouds are actually electrified and that the lightning discharge is an electrical phenomenon. No claim is made that Diviš performed such an experiment himself. Hence we are probably justified in assuming that he was aware of such experiments made by others. Indeed, Hujer quotes a letter, dated 26 July 1753, written by Fricker (who had visited Diviš), which refers to the "iron rod" used by Richmann to conduct "atmospheric electricity"[7]; very likely, Fricker would have informed Diviš (or would have discussed with him) these Russian experiments on atmospheric electricity.

Hujer presents evidence that Diviš was not completely isolated. He went to Vienna in 1750, he was in contact with various members of the court, he was a correspondent of Scrinci (professor of physics in the university at Prague), and was a friend and correspondent of K. W. Oetinger at Waldorf near Tübingen (who sent his pupil Fricker to visit Diviš in 1753). We are told by Hujer that Diviš "used Leyden jars . . . soon after their . . . discovery." Diviš, in other words, appears to have been informed of developments in the rapidly advancing science of electricity. It seems wholly unlikely that Franklin's work could have wholly escaped him.

Franklin's experiments and ideas were reported by him to Peter Collinson and John Mitchell. Notice of his communications appeared in London magazines and the *Philosophical Transactions* of the Royal Society of London. His papers were collected in book form and published in London in 1751, and this work appeared in French translation in Paris in 1752.[8] A preliminary note about this book appeared in the March 1752 issue of the widely circulated *Journal des Sçavans* and a long review opened the May 1752 issue. This review contained an account of Franklin's views on the electrical nature of the lightning discharge and it also described the lightning rod proposed by Franklin to prevent damage to buildings and ships: "M. Franklin . . . va même jusqu'à donner un moyen de préserver de la foudre les maisons élevées, les églises, les vaisseaux, &c. Ce moyen est fondé sur la force des pointes pour attirer le feu électrique. Il consiste . . ." The French edition of Franklin's book was reviewed in other publications, such as the *Journal œconomique* (November 1752 et seq.) and the "Mémoires de Trévoux" (June 1752).[9] Franklin's book aroused considerable excitement in France, especially after the demonstrations described in it had been performed before the king and court to a general applause. What gave the book its greatest success, however, was the successful execution of the sentry-box experiment, under the direction of Jean-François Dalibard, translator of Franklin's book, during a thunderstorm on 10 May 1752 (see chapter 6). Dalibard read a report of the experiment and the Franklinian principles it demonstrated to the Académie Royale des Sciences on 13 May.[10] Five days later, Delor repeated the experiment with equal success and, as the news spread across Europe, the experiment was repeated—with some variations—in France, in Germany, in Russia, and in Belgium. Reports of these experiments were widely circulated in learned journals, newspapers, magazines, and by word of mouth. Jacques de Romas, "assesseur au présidial de Nérac," who independently conceived the lightning kite experiment, learned of the sentry-box lightning experiment by reading an account in the *Gazette de France* for 27 May 1752.[11] The news traveled to Russia and was reported in the *Saint Petersburg Journal* in 1752 and, during the same year, the lightning rod was mentioned in a poem by Lomonosov "On the Usefulness of Glass."[12] We may note that Franklin's name was usually associated with his discovery and that Dalibard had said explicitly that he was merely following Franklin's instructions.

The account in the *Saint Petersburg Journal* stated, "In Philadelphia, in North America, Mr. Benjamin Franklin has been so daring as to try to extract from the atmosphere that terrible fire which so often destroys vast areas." A review of Winkler's *Programma* (1753) in the Leipzig *Nova acta eruditorum* referred explicitly to Franklin's research, his book, and Nollet's book attacking him.[13] Incidentally Nollet's book, written in opposition to Franklin, and many times reprinted, was an important vehicle for spreading Franklin's ideas.[14] Franklin's work was described in the *Göttingische Anzeigen von gelehrten Sachen,*[15] and G. W. Richmann noted Franklin's priority in an article in the *Novi commentarii Academiae Scientiarum Imperialis Petropolitanae.*[16] The Italians also knew who had invented the lightning rod that they erected.

Although many "electricians" mentioned the names of scientists who had—prior to Franklin—held that lightning is an electric discharge (favorite names were Gray, Hales, Nollet), the published records indicate a general appreciation of the fact that Dalibard's Marly experiment was designed by Franklin, and that Franklin was plainly the first to design an experiment to test the electrification of clouds and to devise an instrument to protect ships and buildings from destruction by lightning. Finally, we may note that the tragic death of Richmann, experimenting on lightning with an ungrounded rod, gave Franklin's experiment a dramatic notoriety. Winkler wrote of this period, "Mr. Benjamin Franklin in the letters he wrote in 1747, 1748, and 1749 from Philadelphia to Mr. Collinson in London, and particularly by the hint in the fourth letter, how to discover by trial whether the matter of electricity and the matter of thunder were really one and the same, gave occasion to these observations. Now what had been observed of the electricity of thunder in 1752 in France, was confirmed in other countries besides by different trials, which M. Mylius has described in the *Physical amusements,* P. 17. After professor Richmann had been killed by a flash of lightening in observing at Petersburg on August 6, 1753, the electrical force in thunder and lightening, I caused to be made an instrument, by which the electrical effects of thunder may be observed at a distance without danger, and I described it in the *Programma de avertendi fulminis artificio ex doctrina electricitatis . . .*"[17]

Diviš knew of Richmann's death before he erected his lightning rod; hence, he knew that lightning is an electrical discharge and, therefore, that lightning must behave like the laboratory discharges, the only difference being one of scale.

Although Professor Hujer suggests that the Continental experiments on the nature of lightning were not always associated with Franklin's name, we have seen that in prominent places, Franklin's name was associated with lightning experiments in France, Germany, Italy, and Russia.[18] It seems inconceivable that Mylius and Ludolf (the account of whose Berlin lightning experiments was sent to the Royal Society of London by Euler) were ignorant of the name of Franklin when they repeated his experiments; even had they been in ignorance at the time of writing, they would quickly have learned whose experiment they had been performing once they saw the published account of their own experimental results in the *Philosophical Transactions.*[19]

Hujer writes that Diviš's plan of his lightning rod was "all complete by the end of 1752 or the beginning of 1753." No evidence is given in favor of the earlier of these dates. By that time, as we have seen, the news of the Marly experiment had traveled widely throughout Europe; Franklin's proposed lightning conductor was being discussed. In 1753 Diviš mentioned his "rod" to Fricker, referring to Richmann's experiments. By the time he erected the rod in 1754, he knew of the Russian experiments. Diviš rod was actually a variation of Richmann's apparatus, notable chiefly for grounding the rod rather than leaving it insulated. Richmann's instrument was only a variation of the insulated test rod designed by Franklin for the sentry-box experiment.

We are told that differences in structure and conception between Diviš's rod and Franklin's adequately demonstrate that Diviš's "machina meteorologica" was "the result of independent search and discovery" rather than depending in any way on Franklin's writings or ideas. It appears that Diviš's rod was a many-pointed multiply grounded conductor designed to "prevent the accumulation of an electric charge or lightning," rather than an instrument to attract and safely conduct a lightning discharge into the ground. Certainly Euler tends to confirm this view in his description; he says that he had received a letter from Diviš assuring him "that he had averted, during a whole summer, every thunder storm which threatened his own habitation and the neighborhood, by means of a machine constructed on the principles of electricity." Euler provides the additional information: "I have been assured that his machine sensibly attracted the clouds, and constrained them to descend quietly in a distillation without any but a very distant thunder-clap."[20] Euler's description sounds very much like Franklin's first idea about the action of a lightning rod. In 1750 he wrote of the grounded pointed iron conductor, that "the electrical fire would, I think, be drawn out of a cloud silently, before it could come near enough to strike . . ." In the first English edition of his book, which was translated into French prior to Diviš's supposed independent invention of the lightning rod, Franklin also says: "Would not these pointed rods probably draw the electrical fire silently out of a cloud before it came nigh enough to strike . . . ?" Franklin also thought that electrified clouds—supposing that thunderclouds were electrified—would be attracted down by the rod but would be discharged of their "electrical fire" before they reached the striking distance. Franklin, who soon realized that grounded conductors would also safely conduct a stroke into the ground, later often complained about those people who objected to lightning rods and who never fully appreciated this second mode of action (which was probably more important, and which had not been contained in his original publication).[21]

The many accounts of experiments with insulated rods in 1752 indicate the widespread belief that an ungrounded rod can "draw off" some of the "electric fluid" in clouds and thereby lessen the probability of a damaging stroke (see chapter 6). If Diviš knew of Franklin's protective rod, which was grounded and which would draw off such electric fluid at a greater rate than an ungrounded rod, then his instrument would be an application of Franklin's first idea on the mode of action of the rod rather than the later one. Thus we would understand why Diviš apparently erected his rod in the garden rather than on the top of a house, since it could draw off electric fire in both circumstances. Dalibard's test rod was likewise erected in a garden. But if Diviš really had confidence in the lightning rod, and if he truly understood that the rod would conduct a stroke of lightning into the ground, why did he not erect his "machina meteorologica" on the highest point possible, that is, atop the house he wished to protect?

Our discussion, while not proving that Diviš was influenced by Franklin's writings, ideas, and experiments, at least makes such an influence seem very likely. In other words, it seems probable that Diviš knew of Franklin's work and then devised a lightning rod of original design, or introduced minor variations. (Diviš invention may have derived from Franklin even though Diviš had never heard his name; the question is not Franklin's reputation, but the spread of his ideas and experiments.)

Even if it is not likely that Diviš conceived of the lightning rod wholly in independence of Franklin, what of the long-standing belief that the rod which he erected in 1754 was the first protective or grounded lightning rod to be erected in Europe? (It followed by at least two years those erected by Franklin in Philadelphia in 1752.) Hujer points out that Poggendorff and others have insisted that the insulated rods used in the performance of the sentry-box experiment, as by Dalibard at Marly-la-ville, were not true lightning rods since they were not grounded. We agree. On the other hand it seems to be the case that protective or grounded lightning rods were actually erected in France in 1752 and possibly in Belgium. We must keep in mind that the published descriptions of the rods erected in 1752 are sometimes incomplete and in most cases fail to specify whether the rods were grounded or insulated (see chapter 6). But, in addition to the ungrounded rods erected specifically for experimental purposes, which the Abbé Nollet properly called "electroscopes," there are two references to protective lightning rods erected in 1752, two years before Diviš's rod. These were published in British magazines and reprinted by Franklin in the *Pennsylvania Gazette*. The first was a dispatch from Brussels dated 3 July, referring to an iron rod erected by one Sieur Torre on the top of his house to prevent a lightning stroke, which had succeeded "beyond his expectation."[22] The second was from Paris, dated 9 November 1752, reporting that several "iron rods" had been erected to preserve houses from "thunder."[23]

In each of these two notes, the important phrase concerns the intent of protecting houses. Franklin's sentry-box experiments, designed to test the electrification of clouds, employed an ungrounded rod; he had stated explicitly that, if experiment did show lightning to be an electrical phenomenon, then rods which were grounded should be erected for protective purposes. Every statement of Franklin's about protective lightning rods insisted that they be grounded. It seems, therefore, that probably the lightning rod erected in Brussels, and certainly the rods referred to in the Paris report, were grounded—as protective lightning rods should be. Hence, even if Diviš should have conceived of his lightning rod wholly independently of Franklin—which the evidence seems to indicate is extremely unlikely—the device he erected in 1754 was hardly the first protective lightning rod to be erected anywhere in Europe.

Diviš's originality seems to be manifested in the peculiar form that his lightning rod took, with its large collection of metallic points. It is somewhat curious to note that Franklin had at one time thought of lightning rods of this form, but later had given up this idea in favor of the simple pointed grounded iron rod. In a letter published in 1750, Franklin wrote:

> There is something however in the experiments of points, sending off, or drawing on, the electrical fire, which has not been fully explained, and which I intend to supply in my next. For the doctrine of *points* is very

> curious, and the effects of them truly wonderfull; and, from what I have observed on experiments, I am of opinion, that houses, ships, and even towns and churches may be effectually secured from the stroke of lightening by their means; for if, instead of the round balls of wood or metal, which are commonly placed on the tops of the weathercocks, vanes or spindles of churches, spires or masts, there should be put a rod of iron 8 or 10 feet in length, sharpen'd gradually to a point like a needle, and gilt to prevent rusting, or divided into a number of points, which would be better—the electrical fire would, I think, be drawn out of a cloud silently, before it could come near enough to strike; only a light would be seen at the point, like the sailors corpusante. This may seem whimsical, but let it pass for the present, until I send the experiments at large.[24]

While Diviš's lightning rod took a form which Franklin had at one time adopted and then abandoned, it does not seem likely that on this score Diviš was influenced in any way by Franklin, since Franklin's description was published only in the *Gentleman's Magazine,* and was not reprinted in Franklin's book on electricity, nor has it ever been integrated into Franklin's collected writings.

We conclude, therefore, that the idea of having a lightning conductor with a large number of points was conceived independently by Diviš, although anticipated by Franklin, but that his claim to having invented the lightning rod independently of Franklin, like his claim to having erected the first lightning rod in Europe, has never been satisfactorily proved.[25]

8

Prejudice against the Introduction of Lightning Rods

"How astonishing is the force of prejudice even in an age of so much knowledge and free inquiry," wrote Professor John Winthrop of Harvard College to Benjamin Franklin on 6 January 1768, after having read in the *Philosophical Transactions* of the Royal Society of London an account of the destructive effects of lightning on St. Bride's steeple. "It is amazing to me, that after the full demonstration you had given, of the identity of lightning and of electricity, and the power of metalline conductors, they should ever think of repairing that steeple without such conductors."[1]

All too often, we tend to think of the history of scientific thought as an orderly process, one in which proofs are convincing and have the effect of rapidly altering men's minds. The history of science and the history of technology both show, however, that there is always a considerable inertia to the human mind which tends to prevent the acceptance of new ideas or the introduction of new instruments. In 1752 experiment confirmed the hypothesis of "the identity of lightning and of electricity," with the attendant corollary that lightning would therefore be attracted by pointed metallic conductors and could be safely conducted into the ground. Looking backward, we are apt to suppose—as Winthrop did almost two centuries ago—that the new invention would have been given a warm reception and that thenceforth people all over the world no longer need have trembled with fear at the onset of a thunderstorm. While a number of lightning rods were introduced in various parts of the world in the late eighteenth century, the record of history provides a large number of instances of refusal to make use of this new device to prevent the destruction caused by lightning.

It is instructive for us to review some aspects of the opposition to lightning rods, because we may see in the very variety of reasons advanced against their use the normal resistance to change which is characteristic of the human mind. We shall observe the ways in which prejudice tends to prevent radical departures from accepted patterns of behavior, however convincing the evidence may be that the customary practices are useless or even dangerous.

Lightning Rods versus Church Bells

In reply to Winthrop's letter about the force of prejudice preventing the adoption of lightning rods, Franklin explained that to him it did not seem quite so extraordinary that "unlearned men, such as commonly compose our church vestries" should be unacquainted with the benefits of lightning rods, or that they should still be prejudiced against the use of them, "when we see how long even philosophers, men of extensive science and great ingenuity, can hold out against the evidence of new knowledge that does not square with their preconceptions; and how long men can retain a practice that is conformable to their prejudices, and expect a benefit from such practice, though constant experience shows its inutility."[2]

The practice Franklin had in mind was the ringing of church bells during a lightning storm. The practice of ringing church bells to dissipate lightning storms and prevent their deleterious effects had a long tradition in Europe and had been a concomitant to the general belief in the diabolical agency manifested in storms.[3] Many a church bell bore an inscription testifying to the use it might have in dissipating the effects of thunder and lightning. The ritual of Paris for consecrating bells declared that "whensoever this bell shall sound, it shall drive away the malign influences of the assailing spirits, the horror of their apparitions, the rush of whirlwinds, the stroke of lightning, the harm of thunder, the disasters of storms, and all the spirits of the tempest."[4] Typical inscriptions on church bells described their power to "ward off lightning and malignant demons"; stated that "the sound of this bell vanquishes tempests, repels demons, and summons men," or exhorted it to "praise God, put to flight the clouds, affright the demons, and call the people"; or noted that "it is I who dissipate the thunders."[5] A form of inscription, denoting the many uses of bells, has been made famous through Schiller's poem *Die Glocke;* it reads: "Deum laudo, vivos voco, / mortuos plango, fulgura frango."[6]

Despite the sanction given in Catholic countries to the use of bells as a means of preventing the effects of lightning, the fact was notorious by the eighteenth century that many churches were struck by lightning bolts and that many bell ringers were killed during lightning storms.[7] In his reply to Winthrop, Franklin drew special attention to the well-known danger in attempting to use bells to ward off the effects of lightning. Franklin had been more than a little annoyed by a paper of the Abbé Nollet, published in the memoirs of the Royal French Academy of Sciences, in which he had objected to the use of metal conductors outside a building as useless or dangerous. Nollet had given many examples of the effects of lightning in churches (as well as other buildings) in which, as Franklin pointed out, the lightning "was conducted from one part [of the building] to another by wires, gildings, and other pieces of metal that were within, or connected with the building." Among the evidence cited by the Abbé Nollet was a report of M. Deslandes of 1718, describing a storm on the night of 14 April in which lightning had struck twenty-four churches between Landernau and Saint Pol-de-Léon in Brittany; the churches struck by the lightning were the very ones in which the bells had been ringing, while the lightning had apparently spared those churches in which the bells had been silent; one church had been entirely destroyed and two of the four bell ringers had been killed.[8]

Nollet, Franklin wrote, "cautions people not to ring the church bells during a thunder-storm, lest the lightning, in its way to the earth, should be conducted down to them by the bell ropes." Nollet had remarked:

> Bells, by virtue of their benediction, should scatter the thunder-storms and preserve us from strokes of lightning; but the church permits human prudence the choice of the times when it is suitable to use this preservative. I do not know whether sound, considered physically, is capable or not of making a thunder-cloud burst and of causing the discharge of its fire towards terrestial objects; but it is certain and proved by experience that thunder can fall on a bell-tower whether one is ringing it or whether one isn't; and if that happens in the first case, the bell ringers are in great danger, because they pull the cords by which the commotion of the lightning can be communicated to them: it is therefore much wiser to let the bells be silent when a thunder-cloud comes over the church.[9]

Since Nollet had carefully observed the path of lightning when buildings were struck, Franklin found it difficult to understand why he did not advocate the use of lightning rods.[10] For, as Franklin noted, if the lightning on its way to earth could be conducted down the bell ropes, which are bad conductors, would it not be wise to offer the lightning a path through good metal conductors placed outside of the steeple, that is, lightning rods? The lightning would certainly "choose to pass in" the good metal conductor or the rod, "rather than in [the] dry hemp" of the bell rope. The expectation "that the sound of such blessed bells would drive away those storms, and secure our buildings from the stroke of lightning" had not been realized during more than a thousand years of experience, Franklin noted, while "lightning seems to strike steeples of choice, and that at the very time the bells are ringing; yet they still continue to bless the new bells, and jangle the old ones whenever it thunders." One would think, wrote Franklin, "it was now time to try some other trick;—and ours is recommended (whatever this able philosopher [Nollet] may have been told to the contrary) by more than twelve years experience, wherein, among the great numbers of houses furnished with iron rods in North America, not one so guarded has been materially hurt with lightning, and several have been evidently preserved by their means; while a number of houses, churches, barns, ships, &c. in different places, unprovided with rods, have been struck and greatly damaged, demolished, or burnt."

In all probability, Franklin argued, the vestries of "our English churches" were not well acquainted with the facts. Otherwise, as "good Protestants they have no faith in the blessing of bells, [and] they would be less excusable in not providing this other security for their respective churches, and for the good people that may happen to be assembled in them during a tempest, especially as those buildings, from their greater height, are more exposed to the stroke of lightning than our common dwellings." Franklin was being much more generous than the facts in the situation of St. Bride's warranted. When he wrote his letter to Winthrop in 1768, he must surely have known that the spire on St. Bride's Church in London, destroyed by lightning in 1764, had previously been destroyed by the same cause in 1750. The destruction of 1764 was rendered particularly interesting in that it served as the basis of a report by William Watson, published in vol. 54 of the *Philosophical Transactions* of the Royal Society, in which the path of the lightning in St. Bride's was examined in the greatest of detail. In the same volume of the *Philosophical Transactions,* two other papers on lightning appeared; one by E. Delaval, who also reported extensively on the damage caused to St. Bride's, and the other by Benjamin Wilson.[11] While these three authors may have disagreed somewhat on details of construction of the rods, there was no question of their unanimous recommendation of some form of lightning conductors as a preservative for this and similar structures. The conclusion to the affair has been described in Priestley's history of electricity: "My readers at a distance from London will hardly believe me, when I inform them, that the elegant spire which has been the subject of a great part of this section, and which has been twice damaged by lightning (for it is now very probable, that a damage it received in the year 1750, was owing to the same cause) is now repaired, without any metallic conductor, to guard it in case of a third stroke."[12] How could the vestry of the Church of St. Bride's have remained wholly ignorant of the discussion of the fate of their spire by the leading scientific body of the realm?

After the damage to St. Bride's, the Dean and Chapter of St. Paul's asked the Royal Society in March 1769 for advice on protecting their church from lightning. Franklin was a member of the committee, along with Watson, Delaval and Wilson, and John Canton. It was recommended that a form of lightning rod be installed; that is, that a "complete metallic communication [be made] between the cross placed over the lanthorn and the leaden covering of the great dome [which was grounded through the water pipes]; as from its height, if any lightning struck it, it would most probably affect the cross."[13]

It may be thought that this episode marked the end of British ecclesiastical resistance to Franklin's invention. That such was not the case may be seen in the preface (dated 1 May 1843) to the remarkable book on lightning and rods written by William Snow Harris:

> The beautiful spire of St. Martin's church, in London, has been recently rebuilt, at a cost of full one thousand pounds sterling, in consequence of an explosion of lightning, which fell on it in July last. Brixton church, near London, had also to undergo extensive repairs, rendered necessary from the same cause. In January, 1841, the spires of Spitalfields and Streatham churches, were struck by lightning, and the latter nearly destroyed: and in August of the same year an electrical discharge shook the spires of St. Martin's and St. Michael's churches, at Liverpool, both modern edifices of a costly and elaborate construction. In January, 1836, the spire of St. Michael's church, near Cork, was rent by lightning down to its very base; and in the following October the magnificent spire of Christ church, Doncaster, was almost totally destroyed by a similar discharge.
>
> Thus, in the United Kingdom alone, and within the short space of five years, we find at least eight churches to have been either severely damaged or partially demolished by lightning; to this list of casualties may be added the fine old church of Exton, in Rutland, which, according to the public journals, was in great measure destroyed in a thunderstorm, so lately as the 25th of last April. A writer in Nicholson's *Journal of Science,* states that he has made a calculation of the average annual amount of damage done by lightning in England alone, and that it cannot be far short of fifty thousand pounds.[14]

Although opposition did not cease at once, religious prejudice against lightning rods did abate, even in the 18th century. St. Mark's in Venice provides an example. Despite "the angel at its summit and the bells consecrated to ward off the powers of the air," its steeple had been frequently struck and injured or ruined by lightning—in 1388, in 1417, in 1489, in 1548, 1565, 1653, and again in 1745. The tower was badly struck in 1761 and again in the following year. In 1766 (fourteen years after Franklin's discovery of the lightning rod), a lightning rod was finally placed upon it; in 1894 Andrew D. White was pleased to report that "it has never been struck since."[15]

Today, a church with lightning rods upon spire or tower is a familiar sight. Little more than a century ago, however, bells were still being rung in an attempt to prevent the destructive effects of the lightning discharge. One of the most dramatic accounts of a tragic attempt to dissipate a lightning storm is recorded for us by the famous Spanish physiologist Santiago Ramón y Cajal in his autobiography:

> The second event to which I referred, namely, the striking of the school by lightning, with extraordinarily dramatic results, also left a broad stamp upon my memory. For the first time there appeared to me in all its irresistible majesty, that blind and ungovernable force of the universe, which is indifferent to suffering and seems not to distinguish between the innocent and the guilty.
>
> This was the tragic occurrence. We children were assembled one afternoon in the school, engaged in prayers under the leadership of the mistress (the master being confined to bed that day). When the afternoon was already advanced, the sky became rapidly overcast and there were several violent thunder-claps, which did not alarm us; then suddenly, in the midst of the deep abstraction of the prayer, our lips in the act of uttering the words of supplication: "Lord deliver us from all evil," there sounded a terrific crash which shook the building to its foundations, froze the blood in our veins, and cut short abruptly the petition which we had commenced. Dense dust mingled with debris and fragments of plaster dislodged from the ceiling, obscured our eyes, and acrid smell of burned sulphur spread through the place. Terrified and running like mad creatures, half blind with the cloud of dust, tumbling over each other under the shower of falling fragments, we anxiously sought the way out without finding it for a while. More fortunate or less paralyzed by fear than the others, one of the children reached the door and the rest rushed after him in terror. The vivid fear which we felt did not allow us to realize what had occurred. We thought that a mine had exploded, that the house had caved in, that the church had fallen down on to the school, everything occurred to us except a lightning stroke.
>
> Some good women who saw us running distractedly came at once to our assistance, gave us water, cleaned off the dusty perspiration which gave us the appearance of ghosts and bandaged temporarily those who had been hurt. A voice coming from among the crowd called our attention to a strange, blackish figure hanging on the railing of the bell tower. In fact, there, beneath the bell, enveloped in dense smoke, his head hanging over the wall lifelessly, lay the poor priest who had thought that he would be able to ward off the threatening danger by the imprudent tolling of the bell. Several men climbed up to help him and found him with his clothes on fire and with a terrible wound in his neck from which he died a few days later. The bolt had passed through him, mutilating him horribly. In the school, the mistress

> lay senseless upon her dais, also struck by the lightning but without serious injuries.
>
> Little by little we took in what had happened. A bolt or flash of lightning had struck the tower, partly melting the bell and electrocuting the priest; afterwards, continuing its capricious path, it had entered the school through a window, and pierced the ceiling of the lower floor where we children were, shattering a great part of the ceiling, had passed behind the mistress, whom it deprived of sensibility, and, after destroying a picture of the Saviour hanging upon the wall, had disappeared through the floor, by a gap, a sort of mouse hole close to the wall.[16]

There is no way of telling when church bells were last rung for this purpose. Indeed, for all we know, they may still be used in a vain attempt to dissipate storms in some parts of the world. The refusal of the mind to change its ways can hardly be better illustrated than by the slowness with which lightning rods were adapted to ecclesiastical structures in the eighteenth century. It is fantastic to think that in Spain a little more than a century after Franklin's demonstration of the electrical character of the lightning discharge, and the subsequent introduction of the lightning rods, church bells should have still been rung to the grim tune of death to the bell ringer in an attempt to counteract "the powers of the air."

Yet Spain was not alone in continuing this custom at so late a date. A late-nineteenth-century observer reported in the 1860s that in Upper Swabia the bells were still being rung in churches during thunderstorms to drive away the hail and prevent damage by lightning. In some of the churches, there were special bells that were used only for that purpose.[17] At the same period, bells were being rung in Constance during thunderstorms, and eager volunteers always were on hand to assist the sexton despite the fact that over the years many were struck dead.[18] A tourist in the Tyrolese Alps wrote in 1867 that "bell-ringing, as the companion of the thunder-storm, is a permanent institution here. I could not make out whether it was supposed to have a physical influence on the electricity, or to have a propitiary effect in a religious sense, calculated to exempt the district from a calamity."[19] A commentary on the lack of progress is provided in an extract from an issue of the Torquay Directory (twentieth century): "It transpires that in conformity with an old usage, the bells of Dawlish Church were rung during the recent thunderstorms, in the belief that the spirit of the bells would overcome the spirit of the lightning."[20] As late as 1906, an English writer on bells, J. J. Raven, was not ashamed of his belief that there was very likely a rational or natural explanation of the power of bells and that it might indeed be true that the bells can dispel lightning as so many had for so long believed.[21]

A man who knew about lightning, and who was aware that bell-ringers were courting death by ringing the bells during a storm, would find his own feelings of anxiety aggravated rather than soothed by the mournful peal. A graphic description of such a man under those conditions is provided for us by the father of the novelist Fanny Burney in his journal of the tour which he made in Germany, the Netherlands, and the United Provinces in search for materials for the history of music. According to Burney,

> I had been told, that the people of Bavaria were, at least, 300 years behind the rest of Europe in philosophy, and useful knowledge. Nothing can cure them of the folly of ringing the bells whenever it thunders, or persuade them to put up conductors to their public buildings; though the lightning here is so mischievous, that last year, no less than thirteen churches were destroyed by it, in the electorate of Bavaria. The recollection of this, had not the effect of an opiate upon me; the bells in the town of Freising were jingling the whole night, to remind me of their fears, and the real danger I was in. I lay on the mattress, as far as I could from my sword, pistols, watch-chain, and everything that might serve as a conductor. I never was much frightened by lightning before, but I wished for one of Dr. Franklin's beds, suspended by silk cords in the middle of a large room.[22]

Franklin's Lightning Experiments and the Introduction of Lightning Rods

An announcement of the lightning rod, containing directions for constructing one, was published by Franklin in "Poor Richard's Almanack" for 1753[23] An advertisement of this issue of "Poor Richard" appeared in *The Pennsylvania Gazette* for 19 October 1752, no. 1243, the same issue in which Franklin published an account of his famous kite experiment: "In the Press, and speedily to be published, POOR RICHARD'S ALMANACK for the Year 1753."[24] The lightning rod or lightning conductor was to be a long iron rod fastened vertically to the side of a house by iron staples, the upper end—joined to a foot of brass wire ending in a sharp point—rising six to eight feet above the highest part of the building, and the lower end being sunk from three to four feet in the ground. In this description, the action of the rod was mentioned in a statement that a house protected with the new device "will not be damaged by Lightning, it being attracted by the Points, and passing thro the Metal into the Ground without hurting any Thing."

In a letter written from Philadelphia to John Mitchell on 29 April 1749, "Containing observations and suppositions, towards forming a new hypothesis, for explaining the several phenomena of thunder-gusts,"[25] Franklin advanced a number of observations tending to indicate the similarity if not the identity of lightning and electrical phenomena.[26]

In his 1749 paper entitled "Opinions and conjectures, concerning the properties and effects of the electrical matter," which Franklin enclosed in a letter to Peter Collinson written from Philadelphia on 29 July 1750, there was a description of experiments intended to provide a model of cloud discharge.[27] A pair of large brass scales was hung from the cross-beam of a balance by insulating cords of silk. The beam was suspended from the ceiling by a pack-thread, so that the bottom of the scales was about one foot from the floor. The weight of the scales caused the pack-thread to untwist, producing a circular revolution of the scales at the ends of the beam. An iron punch was placed in a vertical position upon the floor, so that the scales would pass over it as they made their circular motion. If one of the brass scales were charged, the system would provide a small-scale mechanical model of an electrified cloud passing over a hill or a high building.

"Now if the fire of electricity and that of lightning be the same," wrote Franklin, "as I have endeavoured to shew at large, in a former paper, this . . . pasteboard tube [in another and similar experiment] and these scales may represent electrified clouds. If a tube of only ten feet long will strike and discharge its fire on the punch at two or three inches distance, an electrified cloud of perhaps 10,000 acres may strike and discharge on the earth at a proportionately greater distance." As the scales moved round it was possible to observe a scale

> draw nigher to the floor, and dip more when it comes over the punch; and if that be placed at a proper distance, the scale will snap and discharge its fire into it. But if a needle be struck on the end of the punch, its point upwards, the scale, instead of drawing nigh to the punch, and snapping, discharges its fire silently through the point, and rises higher from the punch. Nay, even if the needle be placed upon the floor near the punch, its point upward, the end of the punch, though so much higher than the needle, will not attract the scale and receive its fire, for the needle will get it and convey it away, before it comes nigh enough for the punch to act . . .
>
> [Thus we may] see how electrified clouds passing over hills or high buildings at too great a height to strike, may be attracted lower until within their striking distance. And lastly, if a needle fixed on the punch with its point upright, or even on the floor below the punch, will draw the fire from the scale silently at a much greater than the striking distance, and so prevent its descending toward the punch; or if in its course it would have come nigh enough to strike, yet being first deprived of its fire it can not, and the punch is thereby secured from the stroke.

If all these things are actually so, concluded Franklin, then it should be possible to protect buildings and ships by means of upright metallic rods, pointed at the upper end and grounded at the bottom. Franklin assumed that such rods would draw off the "electrical fire" from passing clouds and thereby prevent a stroke.

Franklin evidently had not thought that lightning rods might also safely conduct a stroke to the ground. His explanation of the probable action of lightning rods was based entirely on the property of pointed conductors (especially if grounded) in attracting electrical charge from a body at a considerable distance and, in the case of electrified clouds, silently causing them to discharge their electrical fire before there was any possibility of an actual stroke with its attendant destruction. Franklin soon learned that the lightning rods, in addition to their property of preventing a stroke, would have the second property of conducting a stroke safely into the ground.[28]

To test his conclusions about the electrification of clouds and the consequent electrical nature of the lightning discharge, Franklin designed the sentry-box experiment, described in his book on electricity.[29] At Buffon's urging the book was translated into French by Jean-François Dalibard.[30] Shortly thereafter, this experiment was performed successfully, as reported by Dalibard and by Delor. A report dated 20 May 1752, sent by the Abbé Mazéas of the French Academy of Science to Stephen Hales of the Royal Society, reads in part as follows:

> The Philadelphian experiments, that Mr. Collinson, a member of the Royal Society, was so kind as to communicate to the public, having been universally admired in France, the King desired to see them performed. Wherefore the Duke D'Ayen offer'd his Majesty his country-house at St. Germain, where M. de Lor, master of experimental philosophy, should put those of Philadelphia in execution. His Majesty saw them with great satisfaction, and greatly applauded Messieurs Franklin and Collinson. These applauses of his Majesty having excited in Messieurs de Buffon, D'Alibard, and De Lor, a desire of verifying the conjectures of Mr. Franklin, upon the analogy of thunder and electricity, they prepar'd themselves for makeing the experiments.
>
> M. D'Alibard chose, for this purpose, a garden situated at Marly, where he placed upon an electrical body a pointed bar of iron, of 40 feet high. On the 10 of May, 20 minutes past 2 in the afternoon, a stormy cloud having passed over the place where the bar stood, those, that were appointed to observe it, drew near, and attracted from it sparks of fire, perceiving the same kind of commotions as in the common electrical experiments.
>
> M. De Lor, sensible of the good success of this experiment, resolved to repeat it at his house in the Estrapade at Paris. He raised a bar of iron 99 feet high, placed upon a cake of resin, two feet square, and 3 inches thick. On the 18 of May, between 4 and 5 in the afternoon, a stormy cloud having passed over the bar, where it remain'd half an hour, he drew sparks from the bar. These sparks were like those of a gun, when, in the electrical experi-

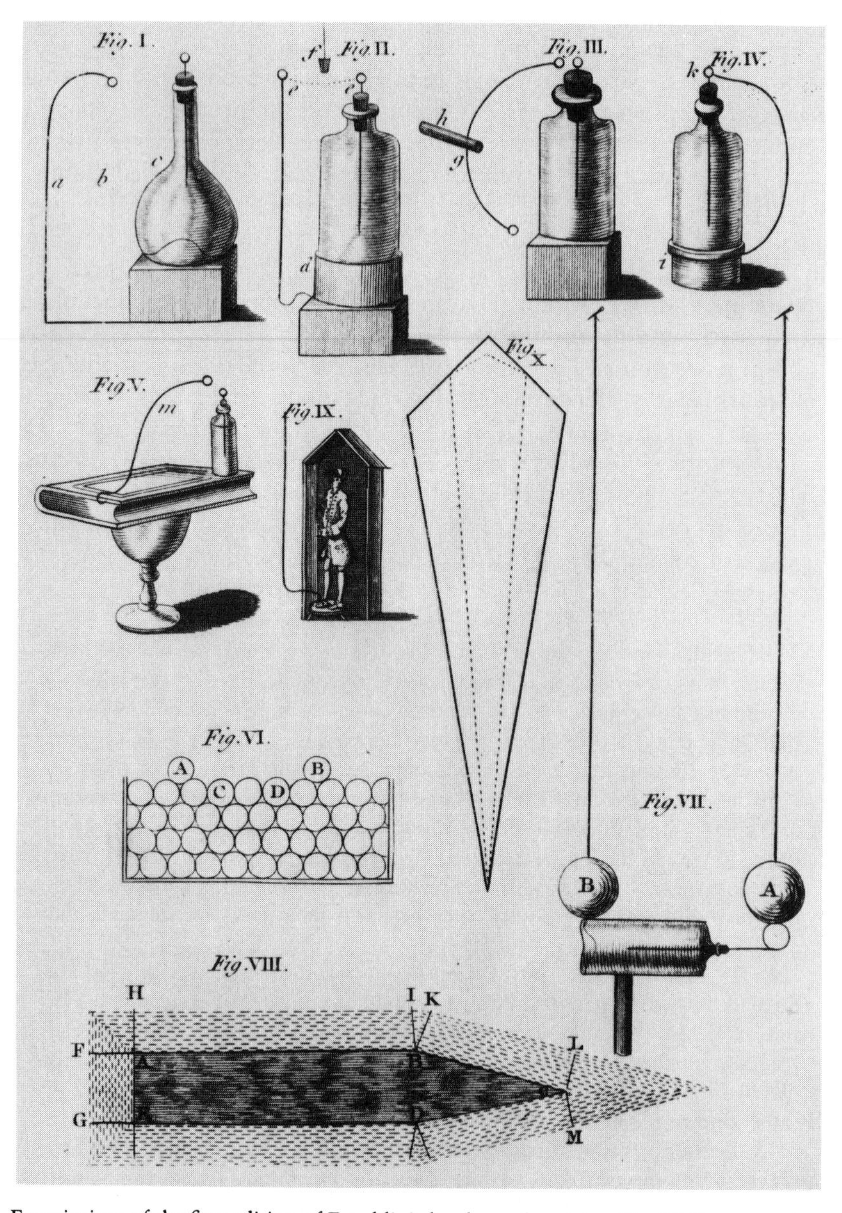

Frontispiece of the first edition of Franklin's book on electricity. Fig. IX illustrates the sentry-box experiment. Burndy Library.

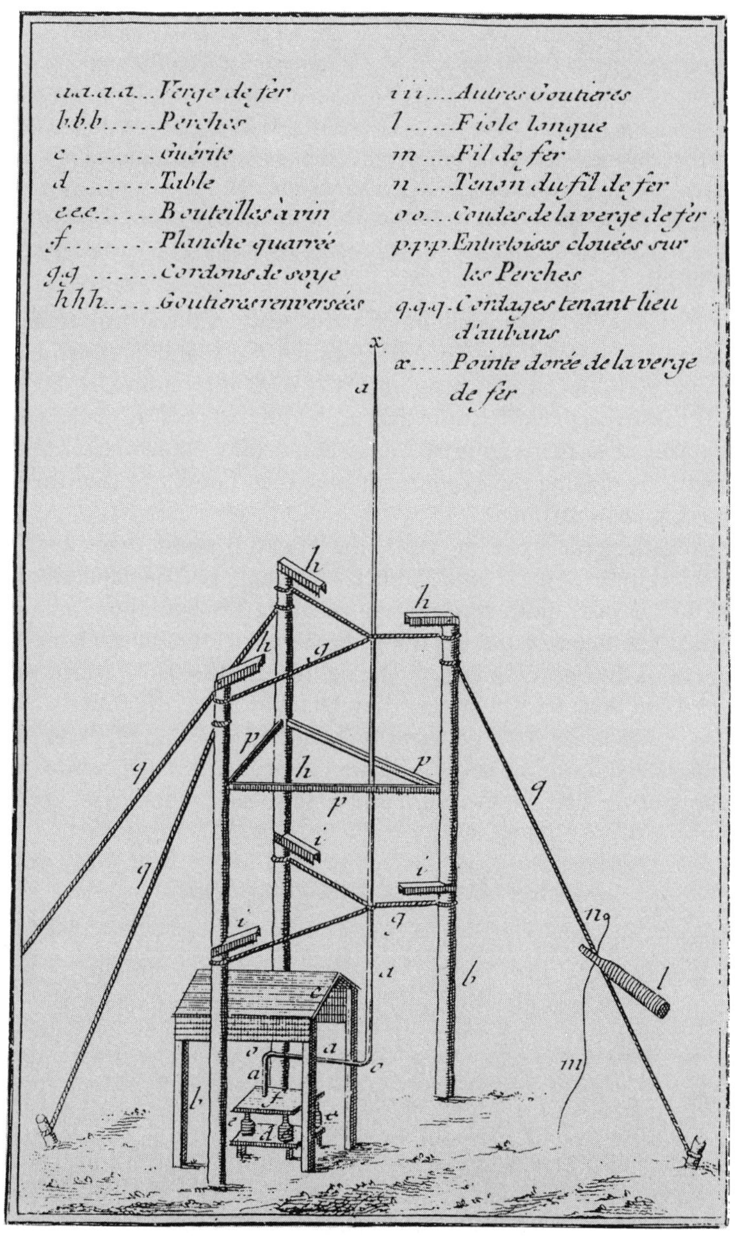

The apparatus for the sentry-box experiment as performed at Marly-la-Ville by Dalibard, from the second edition of Dalibard's translation of Franklin's book. Burndy Library.

> ments, the globe is only rubb'd by the cushion, and they produced the same noise, the same fire, and the same crackling. They drew the strongest sparks at the distance of 9 lines, while the rain, mingled with a little hail, fell from the cloud, without either thunder or lightning; this cloud being, according to all appearance, only the consequence of a storm, which happen'd elsewhere.
>
> From this experiment we conjectur'd, that a bar of iron, placed in a high situation upon an electrical body, might attract the storm, and deprive the cloud of all its thunder. I do not doubt but the Royal Society has directed some of its members to pursue these experiments, and to push this analogy yet further.[31]

On 21 December 1752 William Watson read to the Royal Society an account of successful attempts at repeating the Franklin experiment in England, by John Canton, Benjamin Wilson, and John Bevis.[32] So popular did this experiment become that in the July 1753 issue of the *Gentleman's Magazine* there appeared an account of a "Machine, easily constructed, for making the experiment by which Franklin's new theory of thunder is demonstrated."

During the year 1752 in which the Franklin sentry-box experiment was successfully performed in France, Germany, and England, Benjamin Franklin devised another means of verifying his hypothesis about the electrical character of the lightning discharge. This was the famous experiment of the lightning kite.[33] The lightning kite was described by him in a letter to Peter Collinson dated 19 October 1752; this letter was read at a meeting of the Royal Society of London on 2 December 1752 and was published in the same issue of the *Philosophical Transactions* as the accounts of the French, German, and English sentry-box experiments.[34]

The Action of Lightning Rods

The last paragraph of Franklin's letter of 19 October 1752 on the lightning kite refers to the "points" erected "upon their buildings" by the French, and gives notice that in Philadelphia "points" had earlier been placed "upon our academy and statehouse spires." On 28 September, before writing this letter, Franklin had published in the *Pennsylvania Gazette* a description from Brussels of a lightning rod erected "upon the top of his house" by "the Sieur Torre," who had been able to draw sparks during a storm, apparently repeating in some form Franklin's sentry-box experiment. From the *London Magazine* and the *Gentleman's Magazine* Franklin learned of the erection of "points" in France. Since this information had come to hand by the middle of August, we may conclude that Franklin implied in his letter of 19 October 1752 that the Philadelphia rods had been erected before late July or early August 1752.

Once grounded rods for protective purposes had been erected, it did not take much time for knowledge to be gained that lightning rods could successfully conduct a stroke of lightning into the ground. A natural confusion then arose between the action of ungrounded or test rods and that of grounded or protective rods. The former was a device to attract a little lightning from the clouds in order to perform experiments on it; the latter were designed to discharge electrified clouds or conduct the lightning safely into the ground.[35]

Let us suppose that the rods, being pointed and grounded, did usually attract electrical fire from the clouds and silently conduct it into the ground. Yet, since on occasion they might also be observed to conduct a stroke of lightning into the ground, the conclusion seems perfectly legitimate that the rod which had attracted the fire silently had now invited a stroke of lightning which otherwise might not have come near the building at all. Thus many individuals were opposed to the use of lightning rods because they believed that the rod invited a lightning stroke which might otherwise, with luck, have been avoided entirely.[36]

One of the famous controversies about the nature of lightning rods in the eighteenth century centered on the issue of whether they should end in sharp points (as Franklin had suggested) or round spherical knobs.[37] A considerable part of this controversy hinged on the question of whether they would attract a lightning stroke more readily in the one case than in the other and hence increase the possibility of danger in case the lightning rod would not safely conduct the stroke into the ground. The fear that lightning rods might produce an unnecessary risk has been described as follows:

> Benjamin Wilson . . . [the chief advocate of "blunt" rods or rods ending in metal knobs] . . . was so suspicious of the attractive power of the rod that he went so far as to advise in 1764 that the rod should not extend above the building at all but should end in a blunt point or knob below the roof. He was, however, outdone by others, who considered that the attractive action of the rod could be reversed, and the lightning-stroke repelled, by fitting the tip of the conductor with insulating glass balls. Such balls were fitted to the vane-rod on the steeple of Christ Church, Doncaster. The result was hardly satisfactory, for the steeple was demolished by a lightning-flash not long after.[38]

Wilson, a Fellow of the Royal Society, believed that pointed conductors might promote "the very mischief we mean to prevent" since they act by "soliciting the lightning and . . . also frequently occasioning a discharge when it might otherwise not have happened."

Actually, Wilson was wholly wrong in assuming the attractive influence of a lightning rod to be in any way harmful; "experience over two centuries has shown that by this influence the rod gives complete protection to a building within a certain radius."[39] Yet he was right in his assumption that blunt conductors work just as well as pointed ones, although he did not advance this idea for the proper reason. In experiments performed in the laboratory, where the scale pans of a balance or tufts of cotton serve to represent an electrified cloud, the difference in action between a nearby pointed conductor and a blunt one is marked. But when the actual dimensions of a thundercloud are compared to those of a lightning conductor, there is no difference in action between the two kinds of tip—blunt or pointed. The controversy dissolves in the change of scale from the laboratory to the field.

In a letter to Dalibard dated Philadelphia 29 June 1755, and read at a meeting of the Royal Society of London on 18 December 1755, Franklin wrote that he had been "but partly understood" in the matter of "points in drawing the electrical matter from clouds, and thereby securing buildings." He complained that in Europe only one function of the rods was mentioned, to prevent a stroke; the other function, to conduct a stroke safely into the ground, was ignored.[40] This letter was published in Franklin's book on electricity and in a note added to the fifth edition (1774), Franklin stated:

> Notwithstanding this complaint of the author's, repeated in four editions of his papers, he continues to be misunderstood and misrepresented in this particular, as if he considered it as the sole use of pointed rods, that they might draw off the lightning from the clouds and, prevent a stroke; and the very instances adduced by him to show that the rods had protected houses, by conducting the lightning when there was a stroke, proved by its having melted the points, are said to be instances of conductors being UNSUCCESSFUL,—in a pamphlet entitled Observations Upon Lightning, &c., just published, 1773.[41]

Twenty years after the invention of lightning rods, Franklin was able to report their progress in a letter which he wrote to Professor John Winthrop on 25 July 1773 from London: "Conductors begin to be used here. Many country seats are furnished with them, some churches, the powder magazine at Purfleet, the queen's house in the park . . ."[42] The tone of this letter may be compared to that of his letter to his fellow experimenter Kinnersley eleven years earlier, in 1762: "You seem to think highly of this discovery, as do many others on our side of the water. Here [in England] it is very little regarded; so little, that, though it is now seven or eight years since it was made public, I have not heard of a single house as yet attempted to be secured by it."[43] In 1772, he wrote a report, presumably to H. B. de Saussure, the Swiss naturalist, which described the progress in the use of lightning rods:

> Pointed conductors to secure buildings from Lightning have now been in use near 20 Ye[ar]s in America, and are there become so common, that Numbers of them appear on private Houses in every Street of the principal Towns, besides those on Churches, public Buildings, Magazines of Powder, and Gentlemen's Seats in the Country. Thunder Storms are much more frequent there than in Europe, and hitherto there has been no Instance of a House so guarded being damaged by Lightning; for wherever it has broke over any of them the Point has always receiv'd it, & the Conductor has convey'd it safely into the Earth, of which we have now 5 authentick Instances. Here in England, the Practice has made a slower Progress, Damage by Lightning being less frequent, & People of course less apprehensive of Danger from it; yet besides St. Paul's Church, St. James Church, the Queen's Palace, & Blenheim's House, a Number of private Gentlemen's Seats round the Town are now provided with Conductors, and the Ships bound to the East & W. Indies & the Coast of Guinea begin to supply themselves with Chains for that purpose made by Mr. Nairne, especially since the Return of Messrs Banks & Solander, who relate that their Ship was as they think saved by one of those Chains from Damage when a Dutch Man of War lying near them in the Road of Batavia was almost demolished by the Lightning.[44]

In September 1767, Franklin wrote a description "of lightning, and the method (now used in America) of securing buildings and persons from its mischievous effects":

> An iron rod being placed on the outside of a building from the highest part continued down into the moist earth, in any direction strait or crooked, following the form of the roof or other parts of the building, will receive the lightning at its upper end, attracting it so as to prevent it's striking any other part; and, affording it a good conveyance into the earth, will prevent its damaging any part of the building.
>
> A small quantity of metal is found able to conduct a great quantity of the fluid. A wire no bigger than a goose quill, has been known to conduct (with safety to the building as far as the wire was continued) a quantity of lightning that did prodigious damage both above and below it; and probably larger rods are not necessary, though it is common in America, to make them of half an inch, some of three quarters, or an inch diameter.
>
> The rod may be fastened to the wall, chimney, &c. with staples of iron . . . The lightning will not leave the rod (a good conductor) to pass into the wall (a bad conductor), through these staples . . . It would rather, if any were in the wall, pass out of it into the rod to get more readily by that conductor into the earth.
>
> If the building be very large and extensive, two or more rods may be placed at different parts, for greater security.
>
> Small ragged parts of clouds suspended in the air between the great body of clouds and the earth (like leaf gold in electrical experiments), often serve as partial conductors for the lightning, which proceeds from one of them to

another, and by their help comes within the striking distance to the earth or a building. It therefore strikes through those conductors a building that would otherwise be out of the striking distance.

Long sharp points communicating with the earth, and presented to such parts of clouds, drawing silently from them the fluid they are charged with, they are then attracted to the cloud, and may leave the distance so great as to be beyond the reach of striking.

It is therefore that we elevate the upper end of the rod six or eight feet above the highest part of the building, tapering it gradually to a fine sharp point, which is gilt to prevent its rusting.

Thus the pointed rod either prevents a stroke from the cloud, or, if a stroke is made, conducts it to the earth with safety to the building.

The lower end of the rod should enter the earth so deep as to come at the moist part, perhaps two or three feet; and if bent when under the surface so as to go in a horizontal line six or eight feet from the wall, and then bent again downwards three or four feet, it will prevent damage to any of the stones of the foundation.

A person apprehensive of danger from lightning, happening during the time of thunder to be in a house not so secured, will do well to avoid sitting near the chimney, near a looking-glass, of any gilt pictures or wainscot; the safest place is in the middle of the room, (so it be not under a metal lustre suspended by a chain) sitting in one chair and laying the feet up in another. It is still safer to bring two or three mattrasses or beds into the middle of the room, and folding them up double place the chair upon them; for they not being so good conductors as the walls, the lightning will not chuse an interrupted course through the air of the room and the bedding, when it can go through a continued better conductor the wall. But where it can be had, a hammock or swinging bed, suspended by silk cords equally distant from the walls on every side, and from the ceiling and floor above and below, affords the safest situation a person can have in any room whatever; and what indeed may be deemed quite free from danger of any stroke by lightning.[45]

The final sentence describes the lightning bed to which, as we have seen, Dr. Burney referred in his story about the lightning storm in Bavaria.

I have referred above to the fact that the Abbé Nollet did not endorse the use of lightning rods and have mentioned Franklin's annoyance at Nollet's rejection of the new invention despite his recognition of the nature of the path of lightning. Nollet's attitude was part of his general antipathy to Franklin, whose work had achieved so great a fame and popularity in France that his own position as the leader of electrical thought had been greatly damaged. Furthermore, Franklin's unitary theory of electrical action in terms of the simple movement of a single electrical fluid constituted a flat contradiction of everything Nollet had been teaching on electricity.

In Nollet's report on lightning and lightning rods, published in 1764, and in which he had declared that the rods placed outside of a building are useless and dangerous, he had said, "Our curiosity can perhaps be applauded for the researches which it has caused us to make on the nature of thunder and on the mechanism of its principal effects," but how much more worthwhile it would be if "we could find some means to protect ourselves from them."[46] Such a discovery is supposed to have been made, he went on

> but unfortunately twelve years of proofs and a little reflection teaches us that we can not count on the promises which have been made to us. I have said a long time ago and with regret that all these points of iron which one shoots up in the air, whether as electroscopes, whether as preservatives, can well inform us that it is thundering or that it is about to thunder; but if it is a question of the good that they can do us, I believe that they are more suitable to attract the fire of thunder to us than to preserve us from it: this is sufficiently proved by the death of the unfortunate M. Richmann, whatever way he had used the iron rod of his fatal experiment; I persist in saying that the project of emptying a thunder cloud of the fire with which it is charged is not that of a physicist.

Richmann's electrocution in 1753 seemed to provide an empirical basis for those who feared the lightning rod.[47] Yet Nollet was plainly mistaken in introducing this event into the discussion, since Richmann had been killed while using an ungrounded rod for test purposes and not a grounded rod such as would be used to protect a house. Nollet's introduction of the phrase "whatever way he had used the iron rod in his fatal experiment" (de quelque façon qu'il ait employé la barre de fer de sa fatale expérience) suggests a willful confusion rather than a naive lack of understanding of the role of grounding in experiments with conductors.

This perversion of the facts (or possible—though doubtful—misunderstanding of them) occurred also in Nollet's book *Letters on Electricity,* published in 1753, largely devoted to replying to and attacking Franklin's book on electricity.[48] Although Nollet stated that the letters were intended "less as a criticism of M. Franklin's doctrine than as a defense of my own," the text itself indicates the contrary to be more nearly the case. Nollet had been induced to

> examine with care what may truly be concluded from the experiments proposed by M. Franklin and since carried into execution in France and elsewhere, in relation to the electricity of the clouds during a storm . . . [By weighing every circumstance] and comparing the greatness of the effects, which have been had in view, with the more than apparent insufficiency of the means, which have been employed to produce them, I believe I have seen clearly that considering the electrification of pointed bodies as a proof of lessening the matter of thunder, is abusing a real discovery to flatter our-

> selves with a vain hope; and it is in part to dissipate this error, if it yet subsists, that determined me to print, in these letters, some reflections which I had made at first only for myself, and a few persons, to whom I was desirous of communicating my opinion.

In the first of the nine letters comprising the book, Nollet wrote that "from the phenomenon of Marly-la-Ville, and those discoveries which have been made since and to which it gave rise, have been drawn the two following consequences: one, that the matter of thunder and that of electricity are one and the same; the other, that by the means of pointed iron rods, one might, without noise or harm, draw off all the fulminating matter from a storm-cloud." In Watson's review of this book, in the *Philosophical Transactions* of the Royal Society, he noted at this point: "But our author has shown, that bodies being pointed are not absolutely necessary; and is desirous, we should not too hastily believe, that mischiefs arising from thunder may be averted by the apparatus proposed. He thinks the means vastly too small for the greatness of the cause."

The seventh letter of the book dealt with "the analogy between lightning and electricity." Nollet stated emphatically that he could not believe in the efficacy of lightning rods: "First, because I see too great a disproportion between the effect and the cause; secondly, because the principle, which is given us to support this opinion, does not seem to me sufficiently established." Nollet could not believe that the fulminating matter contained in a cloud capable of covering a great city could be drawn off in a few minutes by a pointed bar as thick as a finger, and ridiculed the credulity of any one who gave serious consideration to lightning rods. Was not this supposed invention to be compared to placing tiny tubes in torrents to prevent a flood? If all that was needed to protect us from the thunderbolts was some pointed bodies placed upon the tops of buildings, would not the spires and crosses (of which the arms usually are pointed at the ends) at the tops of our steeples have been sufficient to provide the protection we desire? Yet churches have never been exempted from the mischiefs of thunder. Thus, as Watson noted, the Abbé Nollet "despairs of our weak efforts ever being able to disarm the heavens." As Franklin pointed out in a great many of his pieces on the effects of lighting, the answer to Nollet lay in the fact that the spires, crosses, and other metal objects on steeples were not grounded by a heavy wire, and might be compared to a test rod but not to a protective one. Indeed, it was this very lack of grounding which made them more dangerous than a mass of metal on top of an ordinary building without a steeple. For this reason the Royal Society's committee had recommended the grounding of the cross atop St. Paul's.

Nollet's objections seem to be based on scientific concerns. Yet one has the feeling, in reading the whole literature of controversy written by Nollet against Franklin's invention, that what really rankled was the wound to his pride in the degree to which Franklin's ideas had superseded his own, and that he objected first and then sought for a scientific basis for rejecting the lightning rod.[49] That his influence considerably retarded the adoption of lightning rods, there can be little doubt.

Two Popes and Two Priests Attempt to Introduce Lightning Rods

All too often we tend to think of churches as vested interests opposing all forms of progress. In the case of lightning rods this is not entirely the case, although it did take some years for Protestants and Catholics alike to begin to erect lightning rods on their churches. Even though the ringing of church bells during lightning storms continued in Catholic countries long after the invention of the lightning rod, it was by no means the case that the Church as an institution was opposed to the new invention. For example, as we saw in an earlier section, it was only fourteen years after the demonstration that lightning is electrical that lightning conductors were placed upon St. Mark's in Venice. Furthermore, in at least two instances that I have found in the eighteenth-century, Church officials tried to erect lightning rods but were frustrated by the antagonism engendered by popular superstition. Despite the enlightened attitude of these Church officials, ignorance prevailed, and in both of these instances the lightning rods already erected were dismantled or destroyed.

The first of these was a lightning rod erected on 15 June 1754, by Procopius Diviš, an amateur "electrician," in the town of Přímětice in Moravia. Diviš was a monk of the Premonstratensian order. His lightning rod—or "machina meteorologica," as he called it—was not like Franklin's and was composed of a pole topped by an iron rod which supported twelve branches of iron curving up like the branches of a tree, each of which terminated in an iron box filled with iron ore, closed with a boxwood cover containing twenty-seven sharp iron points which plunged at their base into the ore. The machine was grounded by metal chains.[50]

The claim is often made that this form of lightning rod was a wholly independent invention on the part of Diviš.[51] There is certainly no ground for doubting that this peculiar form of lightning rod was invented by Diviš; at least no one has ever found a reference to any lightning rod like it prior to 1754. It is, however, difficult to suppose that by 1754 Diviš had not heard of Franklin's experiments to prove the electrical character of the lightning discharge (see Chapter 7).

In any event, Diviš's rod remained standing for six years, until March 1760, when the people of the village tore it down. Apparently the initial cause of hostility was a great drought which the peasants attributed to the malign influence of the rod. It has also been stated that Diviš's enemies, "jealous of his success at the Court of Vienna," excited "the peasants of the locality against him . . . under the pretext that his lightning rod was the cause of the great drought."[52] Whether the peasants' reaction against the rod was spontaneous, or whether the fires of their hatred had been fanned by the enemies of Diviš, the fact remains that the peasants were hostile to the new instrument, and in the end their superstition proved to be sufficiently powerful to effect the destruction of the rod.

A somewhat similar episode occurred in Italy, and is rendered all the more interesting in that the Pope himself interceded in favor of the new invention. Franklin's ideas had been publicized in Italy by Father Giovanni Battista Beccaria; Franklin was partly responsible for having his famous book, *Artificial and Natural Electricity,* translated into English.[53] As a Franklinist, Beccaria attacked the Abbé Nollet. He repeated many of Franklin's experiments—in particular, that of the lightning kite—and became one of the foremost authorities on the subject of atmospheric electricity. Not quite so well known, but of great interest, are Beccaria's experiments on the electrolysis of metallic compounds, about a half a century before the famous experiments of a similar kind by Davy. Performed long before Volta's invention of the electric battery, these interesting electrolysis experiments of Beccaria never attracted in their own lifetime the attention they deserved and even today are largely unknown to chemists.

An important source of information concerning Beccaria is Charles Burney, who published in 1771 the journal of his tour through France and Italy which he had undertaken in order to collect materials for his history of music. Burney had a letter to Signor Baretti of Turin and the latter introduced Burney to Beccaria, "for whom, at first sight, I conceived the highest regard and veneration." Burney described Beccaria as follows:

> He is not above forty; with a large and noble figure, he has something open, natural, intelligent, and benevolent in his countenance, that immediately captivates. We had much conversation concerning electricity, Dr. Franklin, Dr. Priestley, and others. He was pleased to make me a present, finding me an amateur (which should always be translated "a dabbler") of his last book, and the syllabus of the Memoire he lately sent to our Royal Society . . . I left my new acquaintance, impressed with the highest respect and affection for him. I must just mention one particular more relative to this

> great and good man, which I had from Signor Baretti; that he, through choice, lives up six pair of stairs, among his observatories, machines, and mathematical instruments; and there does everything for himself, even to making his bed, and dressing his dinner.[54]

Armed with a recommendation from Beccaria, Burney paid a visit to the Physical Institute at the University of Bologna to see "the Dottoressa Madame Laura Bassi, and met with a very polite and easy reception." Laura Bassi occupies a singular place in the history of eighteenth-century education, when the presence of women on university faculties was rare.[55] Burney described her as "between fifty and sixty; but though learned, and a genius, not at all masculine or assuming." She and Burney talked about the celebrated men of science in Europe, and in particular she expressed great admiration for the English, "in Eulogiums of Newton, Halley, Bradley, Franklin, and others." According to Burney,

> She told me that Signor Bassi, her husband, immediately after Dr. Franklin had proved the identity of electrical fire and lightning, and published his method of preserving buildings from the effects of it, by iron rods, had caused conductors to be erected at the Institute; but that the people of Bologna were so afraid of the rods, believing they would bring the lightning upon them, instead of the contrary, that he was forced to take them down. Benedict XIV, one of the most enlightened and enlarged of the Popes, a native, and in a particular manner the patron, as well as sovereign of Bologna, wrote a letter to recommend the use of these conductors; but it was so much against the inclination of the inhabitants of this city, that Signor Bassi desisted entirely, and they have never since that time been used here.[56]

From what we know of the life and personality of Pope Benedict XIV, his action as described by Burney is wholly in character.[57] No such letter is to be found in the published correspondence of Benedict XIV, although it may well exist in manuscript, and may have been published in contemporary newspapers or journals.[58]

Incidentally, the husband of Laura Bassi was not "Signor Bassi," as Burney supposed, but rather Giuseppe Veratti. The latter repeated Franklin's sentry-box experiment and is said to have "obtained the electric spark in all weathers, through a bar of iron resting in sulphur";[59] his published works deal with the application of the electric shock to paralytics and the aurora borealis, as well as his investigations of atmospheric electricity.[60]

No discussion of eighteenth-century Italian churchmen and lightning would be complete without mention of Abbé Giuseppe Toaldo, professor of physics at Padua, who introduced the lightning rod into the Venetian Republic. Toaldo published a number of works on lightning and lightning rods and translated several items into Italian, including de Saussure's *Exposition abrégée . . .* of 1771.[61] The lightning rod installation in the beautiful medieval cathedral at Siena (whose tower had been repeatedly struck by lightning) was supervised by Toaldo and, although the local inhabitants regarded the rod with terror and apprehension, it remained in place; on 10 April 1777, the rod carried off a heavy stroke of lightning "without doing the slightest damage even to the gilded ornaments near which it passed. The inhabitants now began to look on the heretic rod with more confidence; and it is an important fact," as Snow Harris reported almost fifty years later, "that this church does not seem to have suffered from lightning since."[62]

Despite the demonstrated success of the lightning rod erected on the Siena cathedral, the churches and cathedrals in Italy did not generally gain protection from heavenly thunderbolts. In 1791, the basilica near Assisi which contains the little chapel of the Portiuncula of Saint Francis was struck by lightning and had to be rebuilt. The funds for the restoration were made available by Pope Pius VI, who not only restored the structure to its original condition but also saw to it that "electrical Franklin rods" would be installed so that in the future the basilica would be protected from any further destruction by lightning. This event is commemorated in a Latin inscription on the wall:

> Portiunculae basilicam hanc, Ordinis Minorum matrem, a divo pontifice Pio V aedificatam, VI Kal. Novem. An. MDCCXCI in summo tholi a fulgure percussam, Pius VI P.M., pecunia suppeditata, restaurari et electricis Franklinii virgis ad futuram tutelam muniri iussit.

This may be translated as follows:

> This basilica of the Portiuncula, the mother of the Franciscan order, erected by Pope St. Pius V, was struck by lightning on its dome on Oct. 27, 1791. Pope Pius VI, providing the funds, had it rebuilt and equipped with electrical Franklin rods for its future protection.

So far as I am aware, Pius VI was the first pope to be responsible for equipping a church with lightning rods.

The foregoing information about Benedict XIV and Pius VI, like the fact of the introduction of lightning rods on some Catholic churches in the eighteenth-century, clearly proves that the slowness in adopting Franklin's invention for the protection of churches did not proceed from ecclesiastical ban or dogma. The hindering factor seems to have been the attitude of individual officials who, like their own parishioners, rejected the lightning rod on the basis of personal prejudice or who failed to introduce "electrical Franklin rods" through ignorance of the new technology. In this respect, they probably were in no way different from their Protestant counterparts in England.

Human Presumption and Divine Providence

While we often refer to the eighteenth-century as "the age of enlightenment," we must remember that this phrase applies to the thought of leaders in the realm of the intellect, but obviously not to the vast majority of people. The prevalence of the custom of ringing church bells during storms, and the kind of antagonism to lightning rods which I discussed in the preceding section, indicate that superstition was still strong in the minds of the unenlightened.[63]

In the volume of letters which the Abbé Nollet published in refutation of Benjamin Franklin's electrical ideas, and in which he expressed himself so forcibly in opposition to the use of lightning rods,[64] he applied himself—as Franklin wrote to his friend Cadwallader Colden on 12 April 1753—"to the superstitious prejudices of the populace, which I think unworthy of a Philosopher."

> He speaks as if he thought it Presumption in man to propose guarding himself against the *Thunders of Heaven!* Surely the Thunder of Heaven is no more supernatural than the Rain, Hail or Sunshine of Heaven, against the Inconvenience of which we guard by Roofs & Shades without Scruple.
>
> But I can now ease the Gentleman of this Apprehension; for by some late Experiments I find, that it is not Lightning from the Clouds that strikes the Earth, but Lightning from the Earth that strikes the Clouds.[65]

The last sentence refers to the experiments Franklin had made late in 1752 and early in 1753 with the aid of a lightning indicator. This instrument was used by Franklin to show that clouds may be sometimes electrified negatively as well as positively, and in fact most often negatively.[66]

According to Franklin's theory, the lightning discharge was nothing other than the establishment of electrical equilibrium by the passage of the electric fluid from a positively charged body (that is, one having an excess of the electrical fluid) to a negatively charged body (that is, one having a deficiency of the electrical fluid). Hence if a cloud were charged negatively, equilibrium (or a state of electrical neutrality) was restored by some electrical fluid passing into it from another cloud (one that was neutral or positively charged) or from the earth. The conclusion from Franklin's theory and his observation that clouds are most often electrified negatively was that the lightning discharge usually must proceed from the earth to the clouds, rather than from the clouds to the earth. This result served as a rationalist basis for dismissing completely any notion of presumption in man's using his ingenuity to circumvent thunderbolts from heaven, since the bolts came from the earth. The issue of presumption, however, was raised in America as well as in Europe and must be given its due place in any consideration of the prejudices against the introduction of lightning rods.

The members of the "Junto," the society founded by Benjamin Franklin in Philadelphia which later became the American Philosophical Society, were led to ask at a meeting on 18 January 1760: "May we Place Rods on our Houses to guard them from Lightening without being guilty of Presumption?" At this time, Franklin himself was in London and had been since 1756, and one may speculate as to whether this question would have been raised had Franklin been in Philadelphia. Be that as it may, the question did come up; and at the meeting of 15 February 1760, the following discussion took place:

> The Company agreed to consider the second Query proposed the 18th of last Month viz Whether we may place iron pointed Rods on our Houses to guard them from Lightening without being guilty of Presumption, and after considering or canvassing the Question they were of Opinion that Rods may be put up to guard our Houses without any Presumption. For tho the rending Peal of Thunder may fill the Minds of the Ignorant with Terror who from an Ignorance of its Cause and their natural Superstition may imagine that it is the immediate Voice of the Almighty and the Streaming Lightening are Bolts launched from his Right Hand and commissioned to execute his Vengeance yet in Reason's Eye Lightening or Thunder is no more an Instrument of Divine Vengeance than any other of the Elements. That it is not always directed to execute divine Wrath appears from hence that it most frequently wastes itself on inanimate Things as Trees, Houses, etc. Indeed the Case is this (as by Experiments appears) it is attracted by Eminences. For wherever a Cloud flying over the Earth which has either more or less of the Electric fluid or Lightening than the Earth, comes so near as to be attracted, a stream of Lightening ensues to restore the Equilibrium, and that stream is conveyed by those Objects which having a Communication with the Earth rise nearest the Clouds. This then being the Case and (thanks to God) it having been discovered that Metal is the best Conveyance and that while it can get that it will follow it without damaging any thing else tho' in the nearest contact with it, so far is it from being Presumption to use this Invention that it appears foolhardiness to neglect it. And if to neglect preventing an Evil when it is in one's power is in some measure to be accessory to the bringing it on, it is hard to say how those persons who are so unfortunate as to have their Houses struck or any of their family hurt by Lightening can acquit themselves of being in some measure guilty. It might be mentioned with what Care we endeavour to guard against the bad Effects of other Elements, what means are used both to prevent & remove Disorders of the Body Plagues & Sickness of every sort, and this without any

> Imputation of Presumption; why then should it be imagined more presumptious in the present Case? But what is said suffices for an answer.[67]

Two aspects of this most reasonable discussion strike the eye of the modern reader. First, there is the well-taken point that the divine wrath would hardly be directed at trees, which are often struck by lightning; second, since there is no longer any objection to curing human ills, or removing sicknesses from the body, why should not people protect themselves from the lightning by using the device which, "thanks to God," it had been vouchsafed to Benjamin Franklin to discover? Far from there being any presumption against God's providence, there was a presumption of guilt against any man who suffered loss or damage to either his house or his family by not taking advantage of the new invention.

One of the participants in the above discussion at the "Junto" was Philip Syng, a silversmith who had collaborated with Benjamin Franklin in the performance of the original experiments. Another of Franklin's collaborators was the Baptist clergyman Ebenezer Kinnersley, whom Franklin had helped to get an appointment as Professor of English and Oratory at the Academy (which later became the University of Pennsylvania).[68] Franklin thought very highly of Kinnersley and published many of his letters in his own book on electricity. The electrical discoveries of Kinnersley are recorded in Priestley's history of electricity and his fame was sufficiently great for his work to have received the highest praise from the Abbé Beccaria. Franklin helped Kinnersley in writing a pair of public lectures on electricity; these Kinnersley delivered in Boston, Newport, and also in St. Johns, Antigua, in the West Indies. Kinnersley announced his first Boston lecture in the *Boston Evening Post* of 7 October 1751; the lectures and demonstrations, given at Faneuil Hall, were evidently very popular since they continued until the end of January 1752.

One item in the second of Kinnersley's pair of lectures is of special interest to us: "Various Representation of Lightning, the Cause and Effect of which will be explained by a more probable Hypothesis than has hitherto appeared, and some useful Instructions given how to avoid the Danger of it: How to secure Houses, Ships, *etc.* from being hurt by its destructive Violence."[69] Those who heard Kinnersley's lectures in Boston, therefore, heard about the lightning rod at the end of 1751 or early in 1752, some months before experiment had shown that Franklin's hypothesis had any validity. The lectures as given in Newport soon after Kinnersley left Boston were evidently identical to those which he had given in Boston; the announcement in the form of a broadside printed by Peter Franklin, dated Newport, 6 March 1752, was merely a reprint of the Boston advertisement.[70]

By the following year, however, a completely new note had been introduced. In a broadside dated "St. John's, April 25, 1753," advertising Kinnersley's lectures in Antigua, the second lecture contained the following items:

> XX. An Experiment, shewing how to preserve Houses, Ships, &c. from being ever Struck by Lightning.
>
> XXI. The endeavouring to guard against Lightning, shewn to be not chargeable with Presumption, nor inconsistent with any of the Principles either of natural or revealed Religion.[71]

By this time, the hypothesis of the electrical nature of the lightning discharge had been confirmed by experiments in France, England, and America, and therefore item 20 is stated in much stronger language. But item 21 dealing with presumption, is a novelty. It obviously would not have been necessary prior to the introduction of lightning rods, since the question of presumption actually arose only after the first use of lightning rods.

The question of presumption was also raised in the South. The *Virginia Gazette* for 17 October 1766 contained an announcement of "A course of experiments, in that instructive and entertaining branch of natural philosophy, called electricity, to be accompanied with lectures on the nature and properties of the electrick fire," to be presented for "the entertainment of the curious" by William Johnson. Johnson's course consisted of two lectures, the first of which was devoted to the principles of electrical science, and the second to lightning. According to the announcement:

> In the second lecture this [electric] fire is shown to be real lightning, by its constantly and invariably producing the same effects as lightning does; in proving and explaining which, most of the effects of lightning will be imitated by the electrick fire, such as killing animals, melting metals, tearing and rending bodies through which it passes; together with many curious experiments, naturally representing the various *phenomena* of thunder storms, accounting for their causes, and explaining their effects. A practical method of preserving ourselves, our houses, and effects, from the destructive violence of lightning, will likewise be shown; the efficacy of which will be demonstrated by such indubitable proofs, from experience, as have not hitherto been exhibited to the world.
>
> Those who desire to have their habitations effectually guarded from the fatal violence of one of the most aweful powers of nature (with which this colony in particular has been often dreadfully visited) may learn, from these lectures and experiments, more of the nature and properties of lightning than has been known to mankind until within these few years. They will, at the same time, have an opportunity of being fully convinced that the method

> proposed for their security, if put in practice, with proper precaution, will be attended with success; and consequently that, instead of having any just objection thereto, from a persuasion of its being presumptuous, we have the utmost reason to bless GOD for a discovery so important and eminently useful. "A prudent man foreseeth the evil, hideth himself; but the simple pass on, and are punished." Prov. xxii 3.
>
> As the knowledge of nature tends to enlarge the human mind, and give us more exalted ideas of the God of Nature, it is presumed that this course will prove to many an agreeable and rational entertainment.[72]

This same announcement was apparently repeated in the issue of the *Virginia Gazette* for 24 October. William Johnson also lectured in Charleston; and an advertisement of his lectures appeared in the *South Carolina and American General Gazette* for 6–13 February 1767.[73]

Benjamin Franklin's South Carolina correspondent John Lining, who made important investigations on human metabolism and who repeated Franklin's lightning kite experiment, became embroiled in a controversy over presumption.[74] We are told that in order to quiet the good men of Charleston who were alarmed at the possibility of incurring the divine wrath as a result of erecting lightning rods, the *South Carolina and American General Gazette* "suggested raising lightning rods to the glory of God."[75]

The Winthrop-Prince Controversy and John Adam's Views

The controversy over presumption that has received the most attention from scholars, and is therefore the most well known, occurred between Professor John Winthrop of Harvard and the Reverend Thomas Prince of Boston. Since this famous debate has been the subject of many studies by historians, there is no need here to do more than indicate some of the major issues.[76]

The quarrel between Winthrop and Prince was occasioned by the earthquake of 1755 and operated on two separate levels—scientific and theological. Prince's sermon *Earthquakes, the Works of God* (printed in Boston in 1755) contained the following statement:

> The more *Points* of *Iron* are erected round the *Earth*, to draw the *Electrical Substance* out of the *Air;* the more the *Earth* must needs be charged with it. And therefore it seems worthy of Consideration, Whether *Any Part* of the *Earth* being fuller of *this terrible Substance,* may not be more exposed to *more shocking Earthquakes.* In *Boston* are more erected than anywhere else in *New England;* and *Boston* seems to be more dreadfully shaken. O! there is no getting out of the mighty Hand of *God!* If we think to avoid it in the *Air,* we cannot in the *Earth:* Yea, it may grow more fatal.[77]

The scientific questions raised by Prince are: Do lightning rods draw the electrical substance out of the air? and will the electrical substance that is accumulated in one part of the earth (say the neighborhood of Boston) tend to produce earthquakes there?

Strange as it may seem to the modern reader, in the eighteenth-century the theory that lightning might be related to earthquakes was scientifically acceptable. For example, in the *Pennsylvania Gazette* for 15 December 1737; Franklin printed an article about earthquakes in which the opinion was endorsed "that the material cause of thunder, lightning, and earthquakes, is one and the same, *viz.* the inflammable breath of the Pyrites which is a substantial sulphur, and takes fire of itself."[78] Adherents to this theory later proved to be among the first to applaud Franklin's notions on the electrical nature of the lightning discharge.

Some ten years after Franklin had published the article on earthquakes, he began his investigations on electricity and—as we saw earlier—in the paper entitled "Opinions and Conjectures" he summed up his earlier conclusions about the nature of lightning, suggested the lightning rod, and proposed the sentry-box experiment to test the hypothesis that the lightning discharge is an electrical phenomenon. This paper was read at a meeting of the Royal Society of London, where—contrary to Franklin's statement in the autobiography—it was not greeted with derision but aroused a considerable amount of interest and discussion.[79] An "Appendix to the *Philosophical Transactions*" was added to the regular publication for 1750; this extra issue (no. 479) consisted of 57 articles, filling 150 pages, all devoted to the single subject of earthquakes, and apparently occasioned by consideration of the series of violent earthquakes felt in England during the spring of 1750. Three of these papers were written by the Reverend William Stukeley; one, which had been read before the Royal Society on 22 March 1749, includes the following statement:

> We had lately a very pretty Discourse read here Nov. 16, 1749, from Mr. *Franklyn* of *Philadelphia,* concerning Thundergusts, Lights, and like Meteors. He well solves them by the Touch of Clouds, rais'd from the Sea (which are non-Electrics), and of Clouds rais'd from Exhalations of the Land (which are electrify'd): That little Snap, which we hear, in our electrical Experiments, when produc'd by a thousand Miles Compass of Clouds, and *that* reechoed from Cloud to Cloud, the Extent of the Firmament, makes that Thunder, which affrightens us. From the same Principle I infer, that, if a non-electric Cloud discharges its Contents, upon any Part of the Earth, when in a high electrify'd State, an Earth quake must necessarily follow.[80]

Similar thoughts to Stukeley's were included in a discussion of earthquakes by Stephen Hales, although he did not mention Benjamin Franklin. It will be noted that Stukeley gave Franklin's hypothesis serious consideration in 1750, long before it had been verified by experiment.

During 1750 and 1751 the London magazines devoted much space to letters on the cause of earthquakes. Some writers, agreeing with Stukeley's position, approved the hypothesis that lightning may cause earthquakes, or at least that the cause of both is the same, whereas others expressed violent disagreement. In the *Gentleman's Magazine* for October 1750 there was published a letter by "E. A." reading in part as follows: "In the philosophical solution which I propose to give of this alarming incident, the reader will find no mention made of any such words as *nitre, sulphur, particles, explosion,* or *electrical fire,* which our modern philosophers make use of on all occasions, in explaining all of the phenomena that appear in the heavens above, or are felt in the earth beneath." In the issue of the same magazine for August 1751, "W. M." objected very strongly to Stukeley's theory, and stated: "I adhere still to the *old creed,* and must believe . . . that the occasion of all earthquakes arises from *fire, air,* or *water,* separately or conjunctly." Since Cave, who published the *Gentleman's Magazine,* was also the publisher of Franklin's book on electricity, it is not surprising to find a note added to Stukeley's letter explaining that he "takes notice of a curious discourse from Mr. *Franklin* of Philadelphia (. . . read at the Royal Society, and since printed by E. Cave) concerning thunder gusts, lightning, the Northern lights, and like meteors; all of which are solved by the doctrine of electricity."

So widespread was the speculation that earthquakes might be explained in terms of the action of electricity that John Fothergill, in his preface to Franklin's book on electricity, felt the necessity of showing that Franklin's ideas about the cause of lightning were based on empirical evidence and were not merely part of the general speculation of the time. He pointed out that "from the similar effects of lightning and electricity," Franklin had been "led to make some probable conjectures on the cause of the former; and, at the same time, to propose some rational experiments in order to secure ourselves, and those things on which its force is often directed, from its pernicious effects." Fothergill emphasized that Franklin's discussion of lightning rested on a sound scientific basis and actually led to a series of proposed experiments whereby the truth of his hypothesis might be tested. Thus Franklin's work was to be contrasted to the unbridled speculations of those who would attribute earthquakes to electricity. "It has, indeed, been of late the fashion to ascribe every grand or unusual operation of nature, such as lightning and earthquakes, to electricity," wrote Fothergill, "not, as one would imagine from the manner of reasoning on these occasions, that the authors of these schemes have discovered any connection betwixt the cause and effect, or saw in what manner they were related; but, as it would seem, merely because they were unacquainted with any other agent, of which it could not positively be said the connection was impossible."[81]

In any event, despite the hostility to the theory that the cause of lightning and earthquakes is the same, the theory was given serious discussion during the 1750s.[82] After the experiments in 1752 had confirmed Franklin's hypothesis about the electrical nature of the lightning discharge, the theory that earthquakes are caused by an accumulation of lightning had been given a kind of modus operandi in the omnipresent electrical fluid; it seemed, therefore, much more worthy of serious consideration than it had been up until that time. Hence, when the Reverend Thomas Prince of Boston published his pamphlet in 1755, a number of scientists endorsed the theory that there is one single cause of lightning and earthquakes; since Franklin's experiments had indicated that the cause of the lightning discharge is the electrification of clouds, earthquakes too must be produced by the action of the electrical fluid.

The first part of Prince's pamphlet of 1755 is a slightly revised version of an earlier discourse of 1727. In the meanwhile, as we just saw, Franklin's electrical discoveries had considerably altered the situation; in 1755 Prince felt it necessary to add an appendix:

> Since my composing the foregoing Discourse, the sagacious Mr. *Franklin*, born and brought up in *Boston*, but now living in *Philadelphia*, has greatly surpriz'd the World with his Discoveries of the *Electrical Substance*, as one great and *main Instrument* of *Lightning* and *Thunder:* not I presume as if it excludes the Operation of *sulphureous, nitrous, mineral, watery* and *airy* Substances; but as a principal Means of exciting them in Action, and of working with them those Effects: and with good Reason suggests, that as this *Electrical Substance* seems to be one of the mightiest Agents we know of among material Substances in this lower World, so extremely subtil as to peirce thro' the most solid Iron with greatest Ease and Rapidity and the *Earth* is the grand Source from whence it rises and to which it returns; it seems very likely that this *Electrical Substance*, with the *Others* mentioned, is a *principal Instrument* in producing *Earthquakes*.[83]

Although Prince seems to have relied on Franklin for support of his ideas, the sentiment was not one which Franklin would have endorsed.

In the extract just quoted, Prince suggested that Franklin's explanation of the cause of lightning merely advanced electricity "as one great and main Instrument of Lightning and Thunder"; as if Franklin had not thereby excluded the operation of other substances which were "sulphureous, nitrous, mineral, watery and airy." Prince was in error, because Franklin certainly believed that the lightning discharge itself was nothing more than the passage of electrical fluid from one cloud to another, from a cloud to the earth, or from the earth to a cloud; thunder, however, as an acoustic phenomenon, would probably have been explained in the terms of the agitation of air particles.

After reading Prince's pamphlet, Winthrop composed a reply in the form of *A Lecture on Earthquakes,* which he read in the chapel of Harvard College on 26 November 1755 and then published. There is no need to go into Winthrop's discussion of the theory of earthquakes here. As a good physical scientist, he abhorred the notion that lightning is a cause of earthquakes. We may confine our attention to the "Appendix, concerning the operation of electrical substances in earthquakes; and the effects of iron points."

Winthrop followed the example set by Fothergill in the latter's preface to Franklin's book on electricity: "Philosophy, like everything else, has had its fashions, and the reigning mode of late has been, to explain everything by *Electricity.*" It was not very long ago that "we were amused with pompous accounts of the wonderful effect of electricity in the practice of physic," electricity having been supposed to be the cure-all for "gout, blindness, deafness, dumbness, and what not! . . . Now it seems, it is to be the cause of earthquakes. . . . It is true, the very ingenious Mr. Franklin of *Philadelphia* has, with singular sagacity, and, in my opinion, with happy success, accounted this way for the phaenomena of thunder and lightning; and has made discoveries upon this subject, which are not only extremely curious in speculation, but of high importance in practice. But this is no argument, that electricity is also the cause of earthquakes."[84] As a sample of Prince's misunderstanding of the subject of electricity, Winthrop quotes the following from page 31 of Prince's tract: "[the] *Electrical Substance* subsists and moves to and fro in *different Parties* or *Collections* in the *Bowels* of the *Earth,* as well as in the *Clouds* of the *Air.* And so waving about in *different Parties* in the *Earth* below, though by *Divine direction,* and *surrounding other Substances,* as in the *Air* above; when a *greater* Party comes within the striking Distance of *another,* a *Shock* is immediately effected: and in *Proportion* to the *Quantities* or those several Parties, and the *other Substances surrounded by them,* is the *Shock* in the *Earth,* either *less* or *greater.*" In reply says Winthrop, "The two cases of lightning and earthquakes are no way parallel; . . . the electric substance, when in the bowels of the earth, is in circumstances essentially different from what it is, when in the clouds of the air; [this] will, I think, plainly appear, by taking a brief view of the known laws of electricity, so far as they can be thought to relate to this subject."[85]

Winthrop's outline of the laws of electricity is of the greatest interest since it is one of the best statements of the principles of electrical knowledge according to Franklin's system of which I know. Winthrop was familiar with Franklin's writings and experiments, many of which he had performed for his classes at Harvard College, later using the electrical apparatus which Franklin had helped the college to obtain.[86] Winthrop was not only a trained scientist, but he knew firsthand the exact experimental basis of Franklin's theoretical speculation. In other words, he was able to read Franklin's statements in their empirical background and to limit properly the extensions and applications to be made of them. Prince, on the other hand, took certain dicta of Franklin out of context, rephrased them in his own words, and then tried to apply them to the theory of earthquakes. Due to the limitations of his empirical experience in the actual performance of laboratory experiments, his applications of Franklin's theory were faulty and he extended the theory in a way that was not in keeping with the Franklinian principles of electrical action.

Jared Eliot wrote to Ezra Stiles on 24 March 1756: "I think Mr. Winthrop has laid Mr. Prince flat on back, and seems to take some pleasure in his mortification. The Professor has in my opinion given the best Summary of Laws of electricity that I have ever seen . . ."[87] In this summary, Winthrop laid stress on the way in which bodies may become charged, and how bodies which are charged may give their charge to other uncharged bodies. He also pointed out in great detail the difference between conductors and nonconductors and the way in which conductors lose their charge when grounded:

> It is easy now to see that, though lightning may be accounted for upon these principles, earthquakes cannot. For clouds, which are collections of watery, that is, of non-electric vapors, being *intirely supported and surrounded by air,* which is an electric *per se,* are capable of having more or less than their natural quantity of the electric substance. And therefore, when a cloud, containing more or less than its natural quantity of the electric substance, approaches the earth; or when two clouds, containing unequal quantities of this substance, approach one another; the consequence will be, a discharge of electric substance from that body which contained the most of it, into that which contained the least, till it becomes equally divided between them; which discharge will be accompanied with a flash of lightning and clap of thunder. But where is the *analogy* between this case in the air, and what may be supposed to pass in the bowels of the earth, to cause earthquakes? To make out any thing like an analogy, we must suppose, *first,* two huge non-electric bodies under ground, which for some earthquakes, must be hundreds, if not thousands, of miles in extent; and, *secondly,* that each of these non-electrics is *intirely supported and surrounded by electrics* per se. The *first* of these suppositions has no difficulty in it; because this terraqueous globe consists almost wholly of non-electrics; but where to find such electric

> supporters for these vast conductors, as are required in the *second,* is a point, I humbly conceive, attended with very great difficulty. And yet, without this it is impossible that either of these non-electrics can have more or less than its own natural quantity of electric substance.

Prince had pointed out that "as an equal distribution of this substance in the clouds produces no lightning, . . . so neither will an equal distribution thereof in the earth produce any concussions there, but an unequal distribution in the earth may cause an earthquake."[88] Winthrop, at the end of his summary of the Franklinian principles of electricity, concluded, "that there can be no 'unequal distribution of this substance in the earth;' and consequently, by this gentleman's own confession in Page 20, 'nothing to produce any concussions there.' " In other words, since the earth is a conductor rather than a nonconductor, Winthrop demonstrated that any excess of electricity "thrown" into the earth at a given place, whether through the action of lightning rods or in any other way, would not remain there; it would instantly "spread itself" so that there would no longer be a concentration of charge in one place. Since there seemed to be no possibility of having unequal distribution of charges ("electric substance") in the earth, there was no way in which electricity accumulating in the earth could cause an earthquake.

Having disposed of Prince's theory that the electric fluid can produce earthquakes, Winthrop next turned to an examination of Prince's statement about the action of lightning rods. Although this part of his attack must seem supererogatory to us, Winthrop felt that if Prince's condemnation of "points" were to go unanswered, the effect would be to "fill with unnecessary terror the minds of many persons." Also, it would tend to discourage the use of "iron points" which were "a means of preventing many of those mischievous and sorrowful accidents, which we have so often seen to follow upon thunder-storms."

Winthrop found six fundamental errors in Prince's discussion. The first two errors relate to the action of points in drawing electrical fluid into the earth. According to Winthrop, Prince was in error in stating that they "draw the electrical substance out of the *air.*" The air is an "electric *per se,*"or nonconductor, and the points draw the electrical fluid only from "non-electrics," or conductors such as metals or clouds. Furthermore, the points do not, as Prince implied, draw electrical fluid to the earth by a "*constant and perpetual action,*" but draw electrical fluid from the clouds only when the clouds have an excess of electrical fluid relative to the earth. Although Winthrop's logic was good, his knowledge was imperfect. Dry air was supposed to be an "electric *per se,*" but moist or humid air was plainly a "non-electric." In addition, a number of experiments made with the sentry-box apparatus designed by Franklin had confirmed the discovery (made by Le Monnier soon after the Marly experiments of May 1752) that a pointed, insulated rod will show signs of electrification even if there are no clouds overhead; the generally accepted conclusion being that such a rod will in fact draw electrical fluid out of the air, as well as out of electrified clouds, though in smaller quantities.[89]

So much for the scientific level of the controversy. Let us now turn to theology. One of the topics of controversy between Winthrop and Prince reminds us of Luther's opinion about the use of bells to ward off the action of demons in the air. Although Luther probably never doubted that storms of various kinds are produced by devils, he apparently "regarded with contempt the idea that the demons were so childish as to be scared by the clang of bells."[90] Winthrop wrote:

> I should think, though with the utmost deference to superior judgments, that the pathetic exclamation, which comes next, might well enough have been spared. "O! there is no getting out of the mighty hand of God!" For I cannot believe, that in the whole town of *Boston,* where so many iron points are erected, there is so much as one person, who is so weak, so ignorant, so foolish, or, to say all in one word, so atheistical, as ever to have entertained a single thought, that it is possible, by the help of a few yards of wire, to "get out of the mighty hand of God."

Prince was more concerned with men's souls than with the principles of science. Yet, since he was sure that Winthrop had misunderstood his aim, he replied in a letter dated 15 January 1752, published in the *Boston Gazette* of 26 January. Prince's design

> was *not* to prove that there *must* be admitted a *newly supposed* material Cause, viz. of the *Electrical Fire or Substance,* as either the *only,* or even the *principal,* or even *any* natural Means or Instruments of *Earthquakes;*—I say, *This* was *not* my Design, nor do I, nor did I assert it: I only supposed, that from Mr. *Franklin's* Discoveries—*It seems very likely,* that *This Electrical Substance, with the others mentioned,* is *a principal Instrument* in *producing Earthquakes.* And upon *this Supposition,* or *supposing it may be* a principal *Instrument* in *Concurrence with others* mentioned; *my main Design* was to shew, that in all its Actions, *this Substance* also must act by Powers derived from and directed by that same omnipresent, perfectly intelligent, spontaneous and almighty Being we call by the Name of God; tho' according to his usual Ways of acting in his present Course of Nature: And by several pleasing Passages in Mr. *Winthrop's* Lecture; I doubt not but that in my asserting the Agency of God in all the Operations of the *Electrical* Substance, I have this learned Philosopher fully concurring with me, and will be ready to make it vastly clearer than I have attempted; and with a brighter Evidence:—But I was going to say, that as Mr. *Professor* has un-

> happily *mistaken me* in *several Parts* or my said *Appendix;* notwithstanding all his *granted Premisses* about *Electrics* and *Non-Electrics;* I cannot yet see that he has clearly proved this Conclusion, *that there can be no unequal Distribution of this Electrical Substance in the Earth or Terraqueous Globe:* And therefore, I should have made some *Remarks* and put some *Questions* to Him, as a free Enquirer after Truth, for the further Elucidation of this curious Matter.

However, concluded Prince, "*for the present,* I had rather be apprehended by his Readers to be *mistaken* in a Point of *Philosophy;* than by entering into a more particular Enquiry into the *natural Causes of Earthquakes,* in this extraordinary Season." Too much attention to the natural causes of earthquakes would divert the public mind from "*Matters* of *infinitely greater Moment,*"[91] that is, the warning from Almighty God to the inhabitants of the sinful city of Boston.[92]

The earthquake which had started the Prince-Winthrop controversy had been very severe. John Adams, recently graduated from Harvard College, recorded in his diary that "we had a very severe shock of an earthquake. It continued near four minutes. I was then at my father's house in Braintree, and awoke out of my sleep in the midst of it. The house seemed to rock and reel and crack, as if it would fall in ruin about us. Chimneys were shattered by it within one mile of my father's house." In John Adams's library, now in the Boston Public Library, there is a copy of Winthrop's lectures on earthquakes which had been presented to young Adams as a gift. In this book, as in many others in his library, Adams wrote a number of comments of his own, which tell us much about contemporary opinion.[93]

After reading Winthrop's statement that he did not believe there was in Boston "one person so weak so ignorant, so foolish, or to say all in one word, so atheistical," to have thought it possible to "get out of the mighty hand of God" by the use of a few yards of wire, Adams wrote:

> This Exclamation was very popular, for the Audience in general like the rest of the Province, consider Thunder and Lightning as well as Earthquakes, only as Judgments, Warnings &c and have no Conception of any Uses they can have in Nature.—I have heard some Persons of the highest Rank among us, say that they really thought, the Erection of Iron Points was an impious Attempt to robb the Almighty of his Thunder, to wrest the Bolt of Vengeance out of his Hand; and others, that Thunder was designed as an Execution upon Criminals, that no Mortal can stay. That the Attempt was foolish as well as impious.—And no Instances, even those of Steeples struck, where Iron Bars have by Accident conveyed the Electricity as far as they reached without Damage, which one would think would force Conviction, have no Weight at all.

On the final blank leaf of Winthrop's lecture, Adams wrote the following thoughts about lightning:

> This Invention of Iron Points to prevent the Danger of Thunder has met with all that Opposition from the Superstitions, Affectations of Piety, and Jealousy of New Inventions, that Inoculation to prevent the Danger of the Small Pox, and all other usefull Discoveries, have met with in all Ages of the World.—I am not able to satisfy myself, whether the very general if not universal Apprehension of Thunder, Earthquakes, Pestilence, Famine &c are designed merely as Punishments of Sins and Warnings to forsake, is natural to Mankind, or whether it was artfully propagated, or whether it was derived from real Revelation.

A postscript noted: "An Imagination that those things are of no Use in Nature but to punish and alarm and arouse Sinners, could not be derived from real Revelation, because it is far from being true, tho few Persons can be persuaded to think so."

We may conclude this section with another extract from Adams's diary, under the date of either the third or fourth of December 1758:

> The other night I happened to be at the Doctor's with Ben Veasey; he began to prate upon the presumption of philosophy in erecting iron rods to draw the lightning from the clouds. His brains were in a ferment, and he railed and foamed against those points and the presumption that erected them, in language taken partly from Scripture and partly from the disputes of tavern philosophy, in as wild, mad a manner as King Lear raves against his daughter's disobedience and ingratitude, and against the meanness of the storm in joining with his daughters against him in Shakespeare's Lear. He talked of presuming upon God, as Peter attempted to walk upon the water; attempting to control the artillery of heaven—an execution that mortal man can't stay—the element of heaven; fire, heat, rain, wind &c.

Conclusion

Looking back on the eighteenth-century prejudices against lightning rods, the twentieth-century reader is all too apt to believe that in this field we have made progress—at the very least, we are apt to feel superior to those who used to ring church bells at the onset of a storm. Yet even today the number of buildings with adequate lightning protection is but a small fraction of those that require rods. Many readers may reply that while the danger from a lightning stroke is possibly very great, the terrors experienced in the past in the face of storms were a result of overexaggeration, since the probability of any given house being struck is so small as to make such an event virtually impossible. The accumulated evidence does not support such a viewpoint.

In 1879 Richard Anderson collected a great mass of statistical information on lightning damage; the list of public buildings alone that had been damaged or destroyed by lightning during the preceding century occupied ten pages of print.[94] In Austria during the eight-year period between 1870 and 1877, 1,702 fires had been reported as caused by lightning out of a total of 40,128; in Bavaria, during the thirty-one years from 1843 to 1873, 1,355 persons had been killed by lightning; in Sweden, during the sixty-two-year period from 1816 to 1877, 860 persons had been reported as killed by lightning; in France, during the thirty-year period from 1834 to 1863, 2,038 persons had been struck dead by lightning; in European Russia (that is, "dans les 49 gouvernements de la Russie Européenne, sans compter la Finlande et les gouvernements du ci-devant royaume de Pologne"), during the five-year period between 1870 and 1874, 2,161 persons had been killed by lightning and 4,192 fires had been reported as caused by lightning; in Prussia, during the nine-year period from 1869 to 1877, 1,004 deaths had been reported as caused by lightning; while in England, it was estimated that the number of persons killed each year by lightning was over one hundred.

In 1950, B. F. J. Schonland, one of the foremost scientists in the field of lightning research, was forced to conclude that "lightning still takes a considerable toll of unprotected property in the country from which the rod came." Lightning was given the sixth place in importance among the causes of fire by the United States National Board of Fire Underwriters in 1921; and the annual direct loss of property due to lightning was estimated as being somewhere between twelve and fifteen million dollars per annum. In the state of Iowa alone, lightning was the cause of 924 fires during the five-year interval between 1919 and 1924. Of these, 95 percent occurred in buildings without lightning-rod protection. (The failure of the rods to afford protection in the other 5 percent may be attributed to faults in installation, a warning that despite the proved efficienty of lightning rods, they need expert care.) The magnitude of the problem is revealed in the following data: there are some sixteen million thunderstorms throughout the whole earth during an average year, or some forty-four thousand per day; since each storm has a duration of about one hour, there are, in different parts of the world, some eighteen hundred storms in progress at any given moment; the number of lightning flashes taking place in any given second is, on the average, about a hundred; a complete flash usually represents a discharge of twenty coulombs, although values as high as one hundred sixty coulombs have been observed; a single stroke usually represents a discharge of from two to ten coulombs, but the most common peak current in the return stroke is thirty thousand amperes, while the greatest value ever observed was two hundred thousand amperes.[95]

Today, almost two and a half centuries after Franklin's invention of the lightning rod, there are still great losses caused by fires of lightning origin. During the three-year period 1985–87 there were 8,459 structures in the United States reported as damaged by fires caused by lightning. These produced 591 injuries to people and resulted in a financial loss of $115,023,000. These numbers come from the Office of Fire Data and Analysis, United States Fire Administration (in the Federal Emergency Management Agency), which estimates that it "receives reports on about 40 percent of the fires reported to fire departments in the United States each year." Presumably the total damage is more than the double of these figures, averaging somewhere between seventy-five and eighty million dollars per annum.[96] Through ignorance, superstition, or carelessness we are annually suffering a tremendous loss that could easily be prevented by Benjamin Franklin's invention.

Our negligence in not erecting lightning rods is hardly due to bravery; most readers probably have at least one member of their family who is to some degree frightened of thunder and lightning. The fright may not be sufficient to make necessary the construction of the Franklin bed for which Dr. Burney yearned while in Bavaria in the eighteenth century, but it may be sufficient to cause severe anxiety during a storm. This kind of anxiety is not always allayed by the presence of a lightning rod; on the contrary, the very sight of a lightning rod—or even the knowledge that a lightning rod is attached to the house—may in itself be a cause for further anxiety of the very same sort that plagued so many people in the eighteenth century, an apprehension that the rod may invite destruction of those whom it is supposed to protect.[97] This fear is, of course, wholly irrational; the evidence accumulated over more than two hundred years proves conclusively that a properly installed lightning rod does not invite destruction, but will provide protection from the lightning.

The anxiety produced by lightning rods is not based on a knowledge of the action of pointed conductors, but seems to be rather a manifestation of a common fear that a dreadful fate lies in store for anyone who dares to tamper with the larger and more potent forces of nature. The idea that man, vis-à-vis the forces of nature, can go just so far and no further, is as old as civilization itself. The taboos of primitive people and the mythology of antiquity alike bear witness to concern about the limits of man's possible activity. We all remember the myth of the flight of Icarus, with the aid of wings fashioned of wax and feathers; he rose up in the air higher than a man may go, and his punishment was that the wings were melted by the sun's heat and he plummeted to earth and was destroyed. Closer to the topic of this essay is the myth of Prometheus, who may be regarded as the founder of civilization; he stole fire from the heavens and bestowed it on man, together with the arts that made its control possible. In anger at this act of presumption, Zeus had Prometheus bound to Mt. Caucasus; among the punishments which ensued was that a vulture came every day to devour his liver, which grew again during the night so that the torture might go on indefinitely.[98]

Franklin, who had presumed to draw down the lightning from the skies, was often likened to Prometheus, as when the German philosopher Immanuel Kant called him "the modern Prometheus," or when Turgot penned his famous epigram, "Eripuit coelo fulmen" (He snatched the lightning from the sky). I do not believe that either Kant or Turgot envisioned a Prometheus-like fate for Franklin. The "enlightenment" of the age enabled them to be somewhat free of the superstitions of the crowd.

As one might expect, the Abbé Nollet did not share in this general feeling of applause at man's willingness to stand up to the lightning. He was frightened at the possible consequences of such a dangerous act as interfering with lightning. Soon after the Marly experiments, he wrote of the anxiety aroused within him by the conduct of his fellow experimenters. It was no longer a joke. "By experiment after experiment," he wrote, "we have succeeded in touching the fire of heaven; but if through ignorance or temerity our profane hands abuse it, we would certainly have cause to repent; and what would we do if some grievous accident would cause us remorse, if consumed by unnecessary regrets, we were to bring into being the Prometheus of the fable and his vulture!"[99]

Nollet's fears seem to have been justified by the horrible death of Richmann one year later. Although Priestley referred, in jest, to the sentiments of the "magnanimous Mr. Boze, who with a truly philosophical heroism, worthy of the renowned Empedocles, said he wished he might die by the electric shock, that the account of his death might furnish an article for the memoirs of the French Academy of Sciences," noting that "it is not given to every electrician to die in so glorious a manner as the justly envied Richman," most of the printed discussion shows a sombre note. Morbid curiosity demanded information on every detail, and the printed accounts lay stress on the rapid decomposition of Richmann's corpse, and the frightful odor arising from it; a quick burial was therefore provided and no autopsy could be made.

We are told that the lightning may represent the punishing power of the father, and that fears of both are closely related.[100] This notion renders especially interesting a statement attributed to the Abbé Nollet, who, as we saw, dreaded the possibility of our realizing the myth of

Prometheus; Nollet's statement reads that it is "as impious to ward off God's lightnings as for a child to resist the chastening rod of the father."[101] Today we would hardly call the child's resistance "impious"; in fact we would more likely ask: why should the child not "resist the chastening rod of the father"? In Franklin's willingness to rise above the superstition of his age, and particularly in the ease with which he ignored the possibility of a Prometheus-like fate and the wrath of the father's rod, we see him as an emancipated spirit and a herald of our modern age.

9

Heat and Color

Of all Franklin's nonelectrical experiments, those which he conducted to determine the relative ability of clothes of different colors to absorb and to conduct solar heat are best known; they have also been subject to much criticism. The usual source of information concerning these experiments is a letter which Franklin wrote to Mary Stevenson on 20 September 1761.[1] This letter was introduced by Franklin into the fourth edition of his *Experiments and Observations on Electricity* and was reprinted in the fifth edition.[2] The portion relating to these experiments reads as follows:

> As to our other Subject, the different Degrees of Heat imbibed from the Sun's Rays by Cloths of different Colours, since I cannot find the Notes of my Experiment to send you, I must give it as well as I can from Memory.
>
> But first let me mention an Experiment you may easily make yourself. Walk but a quarter of an Hour in your Garden when the Sun shines, with a part of your Dress white, and a Part black; then apply your Hand to them alternately, and you will find a very great Difference in their Warmth. The Black will be quite hot to the Touch, the White still cool.
>
> Another. Try to fire Paper with a burning Glass. If it is White, you will not easily burn it; but if you bring the Focus to a black Spot, or upon Letters, written or printed, the Paper will be immediately on Fire under the Letters.
>
> Thus Fuller and Dyers find black Cloths, of equal Thickness with white ones, and hung out equally wet, dry in the Sun much sooner than the white, being more readily heated by the Sun's Rays. It is the same before a Fire; the Heat of which sooner penetrates black Stockings than white ones, and so is apt sooner to burn a Man's Shins. Also Beer much sooner warms in a black Mug set before the Fire, than in a white one, or in a bright Silver Tankard.
>
> My Experiment was this. I took a number of little square Pieces of Broad Cloth from a Taylor's Pattern-Card, of various Colours. There were Black,

> deep Blue, lighter Blue, Green, Purple, Red, Yellow, White, and other Colours, or Shades of Colours. I laid them all out upon the Snow in a bright Sunshiny Morning. In a few Hours (I cannot now be exact as to the Time), the Black, being warm'd most by the Sun, was sunk so low as to be below the Stroke of the Sun's Rays; the dark Blue almost as low, the lighter blue not quite so much as the dark, the other colours less as they were lighter; and the quite White remain'd on the Surface of the Snow, not having entred it at all.
>
> What signifies Philosophy that does not apply to some Use? May we not learn from hence, that black Clothes are not so fit to wear in a hot Sunny Climate or Season, as white ones; because in such Cloaths the Body is more heated by the Sun when we walk abroad, and are at the same time heated by the Exercise, which double Heat is apt to bring on putrid dangerous Fevers? That Soldiers and Seamen, who must march and labour in the Sun, should in the East or West Indies have an Uniform of white? That Summer Hats, for Men or Women, should be white, as repelling that Heat which gives Headachs to many, and to some the fatal Stroke that the French call the *Coup de Soleil?* That the Ladies' Summer Hats, however, should be lined with Black, as not reverberating on their Faces those Rays which are reflected upwards from the Earth or Water? That the putting a white Cap of Paper or Linnen *within* the Crown of a black Hat, as some do, will not keep out the Heat, tho' it would if placed *without?* That Fruit-Walls being black'd may receive so much Heat from the Sun in the Daytime, as to continue warm in some degree thro' the Night, and thereby preserve the Fruit from Frosts, or forward its Growth?—with sundry other particulars of less or greater Importance, that will occur from time to time to attentive Minds?[3]

The experiment described in the second paragraph of the letter was actually performed some years earlier; it became well known in part because when Franklin described it in this letter, he made the oft-quoted statement "What signifies Philosophy [natural philosophy, science] that does not apply to some Use?"

For many years, this letter was the sole source of information concerning Franklin's experiments. No one discovered the original notes of the experiment, to which Franklin refers in the first paragraph quoted above. Nor do his other letters or his autobiography give any clue to when the experiment was made, whether in the 1750s, the 1740s, or even the 1730s.

There are, however, additional manuscript sources that reveal the probable date and the origins of this experiment, together with the role of similar experiments made by Franklin's friend and associate, Joseph Breintnal, an original member of the "Junto," Franklin's self-improvement club. These documents introduce new knowledge concerning the actual scientific activities of the Junto. We may note that although Breintnal was a considerable personage in eighteenth-century Philadelphia, he did not merit inclusion in the *Dictionary of American Biography*.

In one of our many exchanges about Franklin problems, Carl Van Doren called my attention to an item by an unidentified author in the "Miscellaneous Papers" in the Franklin Collection of the American Philosophical Society Library, described in the Calendar as follows: "1736–7. January 25. Experiments with various colors to show which imbide and which repel the sun's rays. Diss. 1 p."[4] From the description, it seemed that this must be the page of "Notes of my Experiments" referred to by Franklin in his letter to Mary Stevenson. And, although this document is not written in Franklin's own hand, it might very well be a copy that he had had made in order to send it on to someone else, perhaps even to Mary Stevenson. This was a practice common enough with Franklin.[5]

The document reads as follows:

Janry 25th 1736/7.

EXPERIMENTS with Colours of various Sorts, to shew which imbibe, and which repel, or do not readily admit, the Sun's Rays; made with placing on the level Snow, Bitts of Linnen, Silk, Leather, Paper, Woollen Cloth, Feathers, & other Materials; exemplify'd by six Degrees of melting down or sinking in the Snow. By which it was observable that the different Wrights or kinds of the Subjects made no Alteration; all the Variety of Effects being owing only to the Difference of Colours, except the piece of Glass.

1st degree	Shallowest. White.
2nd	Less shallow. Light Red. Red & White striped. Light Yellow. Light Azure.
3rd	Least shallow, meaning next above deep. Lively Blood Red. Reddish Brown, or Bright Cinnamon.
4th	Deep. Deep Grass Green. Yellow Brown, or dirty Yellow.
5th	Deeper. Deep Blue. Gloomy Red. Dark Olive, or dark Brown.
6th	Deepest. Black. Also a piece of Window Glass.

Note. Some of the above Colours may be a little misnamed; or if another Man should make the Use of such Colours as he would take for some of the above mentioned, his Experiments might not exactly agree with these, especially in those that are nearest the Middle between the two Extreams of Black and White; therefore it may be proper to note for a Certainty that such Things as are perfectly Black, sink deepest; such as are dark Brown. dark Blue, or otherwise nearest akin to Black, sink the next deepest. And such Things as are most perfectly White sink the least, and such as are of a pale Flame Colour, or that vary the least from Witeness and Light, sink the least next to White.

Memo. An easy Way to make a Couple of Experiments that will be very evident and give Satisfaction, is to take an Inch or two of the Feather end of a black Quill and the like Part of a white Quill; also to black with Ink one End of a Slip of white Paper, and the other End remain white; put these on the Snow where the Sun comes, and in an Hour's Time the Black of each will be sunk below the surface (perhaps ½ an Inch, as the Heat may be) and the White scarce visibly settled nor will the White settle so much in a whole Day.

An Observation may be added That those Things which are most closely woven, or are least porous, do most readily shew the Sun's Effect upon them. Also such as have one Side black and the other Side of a different Colour, will imbibe more Heat with the black Side next to the Sun.[6]

This record of experimentation seems to agree in much detail with the account given in the letter to Mary Stevenson. A further search has shown, however, that it is not a note of Franklin's experiment at all. It is a transcript of an experiment made by Joseph Breintnal. Because this transcript appears to have been in Franklin's possession, one may assume that early in 1737, or soon afterward, Franklin knew of Breintnal's experimental proof of the effect of color on thermal absorption and conduction. If Franklin had actually made this experiment himself, and if he did so independently of Breintnal, then his own experiment would very likely have been made prior to 1737. That this is the case will be seen from an examination of Breintnal's papers, in which we find the original account, from which the document quoted above was transcribed.

Joseph Breintnal and His Experiment

Joseph Breintnal or Breintnall (died 1746) was a friend of Franklin's, an original member of the Junto.[7] Breintnal was associated with Franklin in publishing "The Busy-Body" in Andrew Bradford's newspaper, the *American Weekly Mercury,* and assisted Franklin in various ways when he entered business. For a while Breintnal was Sheriff of Philadelphia county and was secretary of the Library Company of Philadelphia from the date of its organization in 1731 until Breintnal's death in 1746, when Franklin himself took over the office. As secretary of the Library Company, Breintnal became a correspondent of the library's London agent, Peter Collinson, who was chiefly responsible for Franklin's initial interest in electrical research.

Breintnal's correspondence with Collinson reveals a keen interest in natural history. One of his letters, in which he describes an aurora borealis, was introduced by Collinson into a meeting of the Royal Society of London and then printed in the journal published by the Society.[8] In addition, Breintnal conducted a remarkable series of experiments in order to determine some of the properties of heat. Not that these experiments have any remarkable qualities when viewed today; they might be performed by any schoolboy. But, at the time when they were made, there were not many people in the American Colonies who were carrying out original experiments in physics.

The most important of Breintnal's experiments concerns the calorific effects of the sun's rays upon pieces of differently colored cloth placed on the snow. His memorandum concerning the experiment reads *in extenso* as follows:

> August 3rd 1737.
>
> That the heat of the Sun penetrates such Things as are colour'd more than such as are White, will appear by the following Experiments, which I was induced to make about seven Years ago, and lately to repeat, from some Hints given me by Benjn Franklin, and from observing that People who come among us from the warm Islands do most of them wear whitish Cloths, in which I suppose they find themselves cooler than in Others tho' few of them may know the Reason of it; from observing also that a small Glass will not burn white Paper, tho' it easily does it if the Paper be stain'd, and from taking Notice of a young Woman's complaining that her black Golves had burnt her Hands, &c.
>
> January 25, 1736.
>
> Experiments with Colours of various Sorts, to shew which imbibe, and which repel (or do not readily admit) the Sun's Rays; made with placing on the level Snow, Bitts of Linnen, Silks, Leather, Paper, Woollen Cloth, Feathers, and other Materials—exemplify'd by six Degrees of melting down or sinking in the Snow. By which it was observable that the different Weights or Kinds of the Subjects made no Alteration; All the Variety of Effects being owing only to Difference of Colours, except the Piece of Glass, and except the Circumstances of Cloth being closely or loosely wove—

1st Degree.	Shallowest	White.
2.	Less shallow	Light Red. Red & White striped. Light Yellow. Light Azure.
3.	Least shallow, meaning next above deep	Lively Blood Red. Reddish Brown, or bright Cinnamon.
4.	Deep	Deep Grass Green. Yellow Brown, or dirty Yellow.
5.	Deeper	Deep Blue. Gloomy Red. Dark Olive, or Dark Brown.
6.	Deepest	Black. Also a piece of Window Glass.

> *Note*. Some of the above Colours may be a little misnamed; or if another Man should make Use of such Colours as he would take for some of the above mentioned, his Experiments might not exactly agree with these, especially in those that are nearest the Middle between the two Extreams of Black & White; Therefore it may be proper to note for a Certainty, that such Things as are perfectly Black sink deepest; such as are dark Brown, dark Blue, or otherwise nearest akin to Black, sink the next deepest. And such Things as are most perfectly White sink the least, and such as are of a pale Flame Colour, or that vary the least from Whiteness and Light, sink the next least to White.
>
> And those Things which are the most closely woven, or are least porous, do most readily shew the Sun's Effect upon them. Also such as have one Side black and the other Side of a different Colour, will imbibe more Heat with the black Side next to the Sun.
>
> An easy Way to make a Couple of Experiments that will be very evident & give Satisfaction, is to take an Inch or two of the Feather End of a black Quill; and the like Part of a white Quill, also to black with Ink one End of a Slip of white Paper, and let the other End remain white; put these on the Snow where the Sun comes, and in an Hour's Time the Black of each will be sunk below the Surface (perhaps ½ an Inch, as the Heat may be) and the White scarce visibly settled; nor will the White settle so much in a whole Day.[9]

The similarity of this document and the one in Franklin's possession (reproduced in the preceding section) leaves no possible doubt that one derives directly from the other. The experiment is dated January 25, 1736/7 in the one and simply January 25, 1736 in the other. This means that the date is actually January 25, 1737 in the "new style" of reckoning years from January 1 rather than the older English style of reckoning years from March 25.[10] The second document is written in Breintnal's hand, but the other seems in a different hand. Most likely Breintnal had a copy made for Franklin and this copy does not contain the preamble which Breintnal kept for the sake of the record, but merely contains an account of the experiment proper. Although the material contained in both is identical, item for item, the style varies in several details.

Did Franklin or Breintnal First Conceive the Experiment?

In an article on Breintnal, Stanley Bloore asserts that "The memorandum recording the information is dated August 3, 1737, but the experiment had been originally performed seven years previously. It was repeated at the later date as the result of hints given by Franklin."[11] But the text of the memorandum itself is hardly a sufficient warrant for this inference, which would make Breintnal the originator of the idea. The memorandum states simply: "the following Experiments, which I was induced to make about seven Years ago, and lately to repeat, from some hints given me by Benjn Franklin . . ." Because the phrase "and lately to repeat" is parenthetical, it is equally likely that the memorandum implies that Breintnal was induced to make the experiments "about seven Years ago . . . from some Hints given me by Benjn Franklin." He was also induced to make these experiments on reflecting on observations listed in the preamble.

The date when Breintnal made the original experiments would be around 1729. His statement places the date of the experiment soon after the organization of the Junto, of which Breintnal was a charter member. Franklin recommended that at the meetings, "Mr. Breintnal's poem on the Junto be read once a month, and hummed in concert by as many as can hum it."[12]

Franklin dominated the club and wanted its members to discuss literary, philanthropic, philosophical, and scientific problems. Some of the more scientific questions raised for discussion were:

> Is *sound* an entity of body?
> How may the phenomena of vapours be explained?
> How may smoky chimneys be best cured?
> Why does the flame of a candle tend upwards in a spire?
> Whence comes the dew, that stands on the outside of a tankard that has cold water in it in the summer time?[13]

If Breintnal made his first experiment in the early years of the Junto, as he himself declares, one may be as certain as can be that Franklin would have known about it. This would have been the very sort of thing to be discussed at Junto meetings and Breintnal, as a good Junto member, would have been more than pleased to offer the results of his own experiments as a topic of "philosophical" discussion.

A careful comparison of the two texts shows that Franklin's experiments and Breintnal's are not identical. According to Franklin's letter to Mary Stevenson, he used sample pieces of cloth from a tailor's card. Breintnal used pieces of cloth and feathers and pieces of cardboard. Franklin wrote that the experiment in question was one which he had himself performed: it was "My Experiment," "I took a number of . . . Pieces of . . . Cloth . . ." Franklin was usually generous in giving credit to his friends and coworkers for their discoveries, especially in the fourth edition of his book on electricity, in which the letter to Mary Stevenson was published. He introduced footnotes into that edition (the first edition that he pesonally supervised) explaining that such-and-such a discovery had been made by Syng, another by Hopkinson, and others by Kinnersley.[14] Since he did not follow this procedure in the case of Breintnal, one can fairly assume that the original idea was Franklin's. This would support the reading of Breintnal's statement which I have suggested, namely, that Breintnal had been induced to make the experiments "about seven Years ago . . . from some Hints given me by Benjamin Franklin."

A logical picture emerges if we admit the following reconstruction:

1. Sometime about 1729 or 1730 (or perhaps earlier), Franklin was led to make the experiment with sample squares of cloth and discovered that the amount of heat absorbed from the sun's rays and conducted by the cloth was a function of the color.

2. Since Franklin was interested in encouraging the pursuit of science, that is "experimental philosophy," among the members of the newly founded Junto, he suggested to Breintnal (and possibly to other members) by means of "hints" that experiments on absorption of solar heat by variously colored substances might reveal some interesting information.

3. Acting on Franklin's suggestion, Breintnal made some experiments and noted, as Franklin had already discovered, the effect of color.

4. Seven years later, on 25 January 1736/7, either as a result of an additional prodding from Franklin or simply on his own initiative, Breintnal repeated the experiments with greater care and wrote up an account of his findings. This account was rewritten by a copyist and given to Franklin.

Following this reconstruction, the date of Franklin's original experiment can be set around 1729 at the very latest.

Boerhaave's Chemistry and Franklin's Experiment

In Franklin's letter to Mary Stevenson, the experiments are introduced by a discussion of heat and color in general experience: "Thus fullers and dyers find black cloths, of equal thickness with white ones, and hung out equally wet, dry in the sun much sooner than the white, being more readily heated by the sun's rays. It is the same before a fire; the heat of which sooner penetrates black stockings than white ones, and so is apt sooner to burn a man's shins. Also beer much sooner warms in a black mug set before the fire, than in a white one, or in a bright silver tankard."

A similar set of observations occurs in the preamble to Breintnal's report of experiments. These statements very closely resemble some remarks of Hermann Boerhaave. There can be no question of the fact that reading the great Dutch master influenced the thought of Franklin and his associates. Here is what Boerhaave wrote:

> If this fire determined by the sun, be received on the blackest known bodies, its heat will be long retain'd therein; and hence such bodies are the soonest and the strongest heated by the same fire, as also the quickest dried, after having been moisten'd with water; and it may be added, that they also burn by much the readiest; all of which points are confirm'd by daily observations. Let a piece of cloth be hung in the air, open to the sun, one part of it dyed black, another part of a white colour, others of scarlet and diverse other colours; the black part will always be found to heat the most, and the quickest of all; and the others will each heat the more slowly, by how much they reflect the rays more strontly to the eye; thus the white will warm the slowest of them all, and next to that the red, and so of the rest in proportion, as their colour is brighter or weaker. This is well known to the nations who inhabit the hotter climates, where the outer garments, if of a white colour, are found best to preserve the body from the scorching sun; and black ones, on the contrary, to increase the heat. And it has often been observed by makers of woollen cloth, that if at the same time and place they hang out two wet pieces, the one black, the other white, the former will smoak and dry quickly, but the latter retain its water long; and cloths of other colours will dry so much the slower, by how much their colours are the brighter.
>
> It has also been long observed, that all black bodies are sooner kindled and set on flame by the same fire, than those of any other colour . . . If a piece of white paper be laid on the focus of a burning-glass, it will be long before it heat, and very long before it take fire; and as soon as kindled, quits its whiteness, turns brown, and then black; immediately after which it catches flame: whereas, if a black paper be laid on the same focus it immediately takes fire . . .

This extract is taken from Peter Shaw's edition of Boerhaave's *A New Method of Chemistry* . . . as published by himself (1741).[15] This work was called the second edition, but a first edition, made by Peter Shaw and Ephraim Chambers (London, 1727), had been based on a spurious edition. This work, originally printed in 1724, had been edited or written and prepared by Boerhaave's students, who had evidently collated their lecture notes and brought out the book under their teacher's name. Boerhaave responded by preparing a "genuine" version of his own book in a two-volume edition published in Leyden in 1732, each copy of which bore his signature in ink as a sign of its authenticity. It was this genuine edition which served as the basis for the Shaw "second edition" of 1741, and which was also independently translated into English by Timothy Dallowe in 1735.[16]

Franklin's letter to Mary Stevenson in 1761 shows great similarity to the text of the genuine edition of Boerhaave, which we know he had seen in Shaw's English translation. The latter was published in 1741 and was based on the Leyden edition in Latin of 1732, which was reprinted in London in the same year. Hence the suggestion for the Franklin and Breintnal experiments must have come from the Latin version, since the experiments of Breintnal were dated 1737. Breintnal's phrase, "about seven years ago," indicates that he had no record of the exact date of the first experiments, nor did he recall just when they had been performed. Perhaps they had been done five years earlier and not seven.

Of course, Franklin may have been inspired originally by reading the text of the spurious edition in Latin of 1724 or the English version of it by Shaw and Chambers of 1727. But there he would have found a briefer account of this subject, including experiments with only three colors: white, red, and black. In this version, there is no mention of the preference for white clothes in hot weather, nor of the ignition of paper by focusing the sun's rays.[17] Had Franklin read this version of Boerhaave's *Chemistry*, and had he performed his first experiments before encountering the genuine version, he would be entitled to a claim of some originality in this matter—but only to the extent of enlarging the number of colors from the three mentioned by Boerhaave, and the use of tailors' sample cloths. In this case, the repetition of the experiment by Breintnal in 1737 could have been occasioned by Franklin's coming upon the genuine edition in Latin with its fuller discussion of this subject.

Boerhaave's brief account was available in the spurious Latin edition of 1724 and its translation of 1727, the more complete account in the genuine Latin edition of 1732 and the two English versions of 1735 (by Dallowe) and 1741 (by Shaw). Breintnal's 1737 memorandum must have been influenced by the later edition of Boerhaave's book—either through the Dallowe translation or the Latin original, or through suggestions made by Franklin on reading the Latin or Dallowe versions. Franklin's later letter, dated 1761, plainly shows the influence of Shaw's translation of 1741, which we know was in Franklin's hands in 1744. The Library Company had bought the Shaw and Chambers version (1727) in 1732 and the new Shaw version (1741) in 1743–1744.

We still have no evidence for the origin and inspiration of the earliest experiments of Franklin. Breintnal's recollection in 1737 of "about seven years ago" would place Franklin's early experiment in 1729 or 1730, five or six years after the spurious edition of Boerhaave's book and two or three years after its translation. We have no way of being certain, with the present paucity of documents, whether Franklin would then have increased the number of colors and have laid the pieces of cloth on snow, but this is strongly implied by the word "repeat" in Breintnal's memorandum and "my experiment" in the letter to Mary Stevenson. In 1737, when Franklin suggested that Breintnal repeat his experiment, the Dallowe translation had been out for almost two years. Reading this new version might have been the occasion for Franklin's suggestion to Breintnal. It may be noted that Breintnal specifically refers to "a young Woman's complaining that her black gloves had burnt her Hands," an experience mentioned by neither Boerhaave nor Franklin.

Boyle and Newton as Sources of the Experiment

Although Franklin read parts of Shaw's edition of Boerhaave's *Chemistry* when it appeared in 1741, both he and Breintnal would have performed the experiment (1729/30, 1736/37) before that book appeared. Very likely the original hint (1729/30) had come from the spurious edition (1724, 1727) and the suggestion for the later repetition (1736/37) from the genuine edition (1732) or its first translation by Dallowe (1735). The Franklin-Breintnal experiments had two distinct, though related, parts. One was a study of the way in which different colors (and black or white materials) absorb heat; the other a study of the greater ease with which a burning glass will ignite paper that is "stained" (Breintnal) or printed with black ink (Franklin). The subject of igniting paper is mentioned in the introductory part of both accounts. Boerhaave specifically mentions how a white paper put into "the focus of a burning-glass" takes a relatively long time to "take fire," whereas under similar circumstances a "black paper" "takes fire" immediately. This is one of the topics "confirmed" by "experiments."

In his presentation, Boerhaave refers to "the experiments of the Academy *del Cimento,*" who tested various conditions of "Fireing Bodies with a Burning-glass." The experimenters found that white paper and other "white Bodies" could be made to "take Fire" from the action of a "Crystal Lens" but "with more Difficulty than Coloured Bodies." But no specific mention was made of the relative ease of igniting black bodies with a burning glass.[18]

Experiments with white and blackened paper were described in a dramatic manner by British natural philosophers, notably Newton and Boyle. In this context it is significant that Boerhaave's *Chemistry,* studied by Franklin and so many other eighteenth-century physical scientists, reveals an admiration for the British school of natural philosophy. He makes references to the work of Hooke and Halley, and Bacon is cited with veneration. Boyle is quoted again and again, always in terms of the highest admiration, and great respect is paid to Newton's *Opticks.*[19] In this great classic of experiment, Newton discussed the behavior of black and white substances with respect to fire and heat.

Proposition 7 of Book 2 Part 3, of the *Opticks* was intended to show that "the bigness of the component parts of natural bodies may be conjectured by their colours."[20] This proposition is a mixture of observation and speculative hypothesis, in which Newton discussed the size of particles and their "order" in terms of their chemical action and such physical characteristics as density. Thus Newton essayed a discussion of the "bigness of metallick particles," and in general the "thickness" of the particles of matter. Toward the end of this proposition, Newton indicated: "for the production of black, the corpuscles must be less than any of those which exhibit colours. For at all greater sizes there is too much light reflected to constitute this colour. But if they be supposed a little less than is requisite to reflect the white and very faint blue of the first order, they will . . . reflect so very little light as to appear intensely black . . ." In this way we may understand why "fire, and the more subtile dissolver putrefaction, by dividing the particles of substance, turn them to black, why small quantities of black substances impart their colour very freely and intensely to other substances to which they are applied . . ." Most important of all, Newton suggested that consideration of particle size affords an explanation as to "why black substances do soonest of all others become hot in the sun's light and burn (which effect may proceed partly from the multitude of refractions in a little room, and partly from the easy commotion of so very small corpuscles.) . . ."

It should, therefore, be noted that the observed difference in absorption of solar heat between black and white substances, and the ease with which sunlight may ignite black material as compared to white, were discussed by Newton long before Boerhaave. These questions entered Newton's discussion of the fundamental nature of matter, as related to the size of the ultimate particles or corpuscles of which bodies are composed. Since black and white were the extremes, and since the other colors ranged in order between these two extremes, the absorption of solar heat might well be expected to vary more or less continuously through the visible spectrum, but Newton did not say so. Boerhaave's extension of Newton's observations was ingenious, if not particularly astonishing; it may be noted that Newton discussed both the question of the heating of bodies by "the sun's light" as well as the upper limit of heating—that is, actual ignition and burning.

Newton returned to this topic in the Queries at the end of Book 3 of the *Opticks*. In Query 5, he discussed bodies and light acting mutually upon one another in the sense that bodies emit, reflect, refract, and "inflect" (diffract) light and that light heats bodies and puts "their parts into a vibrating motion wherein heat consists." Newton asked (Query 6), "Do not black bodies conceive heat more easily from light than those of other colours do, by reason that the light falling on them is not reflected outwards, but enters the bodies, and is often reflected and refracted within them, until it be stifled and lost?" The "strength and vigor of the action between light and sulphureous bodies observed above" is probably, according to Newton's next Query, one reason why "sulphureous bodies take fire more readily, and burn more vehemently than other bodies do."

Franklin, Boerhaave, and Newton were admirers of Robert Boyle and thus it is of interest to see that Boyle had also explored the problem of the heating of black and white materials by the sun's rays. In his *Experimental History of Colours,* (1663, 1664), Part 2, "Of the nature of Whiteness and Blackness," Boyle described how black and white tiles exposed to sunlight became heated at different rates. He also observed that an egg painted black and exposed to the sun cooked more quickly than a white egg, and that his hand would be warmed more rapidly by sunlight if encased in a thin black glove than in one of "thin but white leather."[21] Undoubtedly, similar observations had been made by others before Boyle.

Hence it may be seen that the experiments described by Franklin and by Boerhaave belong to a tradition of explorations relating to heat and light that goes back well into the seventeenth century. Bacon had discussed the problems of heat and light in relation to solar radiation, and others had examined the possibility of light without heat, as in the case of moonlight. Later on, at the end of the eighteenth century, it was found that radiant heat might be transmitted without being accompanied by visible light.[22] Thus we can understand why, in his letter to Mary Stevenson, Franklin did not say, much less imply, that the effects of color on the absorption of solar heat had been an independent discovery of his—although some later writers have claimed that this should be included among his contributions to physics.[23] Always careful in scientific matters, Franklin merely said that he had performed an experiment, one that we have seen did in fact differ from what had been reported by Boerhaave, if only in details. Perhaps he took it for granted that all persons interested in heat would have been familiar enough with Boerhaave's book and with Newton's *Opticks* to make specific references to them unnecessary. If so, this dramatizes our general lack of familiarity today with what would have been considered the common sources of knowledge of physical scientists of the eighteenth century.

10

The Pennsylvania Hospital

In 1754 Franklin published his book about the Pennsylvania Hospital, a record of its founding and the first years of its operation, intended to elicit support for the new institution. There are many astonishing aspects of this book for today's reader. For example, in the first sentence of the Petition it is stated that the primary reason for a hospital was that "the Number of Lunaticks, or Persons distemper'd in Mind, and deprived of their rational Faculties, hath greatly encreased in this Province." It may seem equally surprising to find (in the third paragraph) a reference to "the Experience of many Years, that above two thirds of the mad People received into *Bethlehem* Hospital [i.e., Bedlam], and there treated properly, have been perfectly cured."

Until 1954 this interesting book had not been reprinted since 1817, nor had it appeared in any of the editions of Franklin's writings.[1] Although it was listed in the standard bibliography of Franklin's writings,[2] neither historians of medicine nor historians of American literature had paid it the attention it deserves. Finally, in 1954, a facsimile edition was issued, and in 1962 the work was included in the Yale edition of *The Papers of Benjamin Franklin*.

From the typographical point of view, the original edition is a splendid example of the good printing done by Benjamin Franklin. Since Franklin's *Account of the Pennsylvania Hospital* was not a record, but a part of his program to gain support for America's first permanent hospital,[3] every device of style, documentation, and format was employed to engage the readers' attention, to hold their interest, and to elicit their sympathetic action. On the last page a form was provided for readers to make contributions, a modern device shrewdly calculated to make the giving of charity easier. I have included here a facsimile of the title page so that today's reader may see how Franklin used the typographer's art to elicit a favorable response. The majestic flowing cadences of the opening

S O M E

A C C O U N T

O F T H E

Pennſylvania Hoſpital;

From its firſt RISE, to the Beginning
of the *Fifth Month*, called *May*, 1754.

P H I L A D E L P H I A:
Printed by B. F R A N K L I N, and D. H A L L. MDCCLIV.

Title page of *Some Account of the Pennsylvania Hospital* (1754). De Golyer Collection of the History of Science, Library of the University of Oklahoma.

paragraph are as impressive today as they were more than two hundred years ago:

> About the End of the Year 1750, some Persons, who had frequent Opportunities of observing the Distress of such distemper'd Poor as from Time to Time came to *Philadelphia,* for the Advice and Assistance of the Physicians and Surgeons of that City; how difficult it was for them to procure suitable Lodgings, and other Conveniences proper for their respective Cases, and how expensive the Providing good and careful Nurses, and other Attendants, for want whereof, many must suffer greatly, and some probably perish, that might otherwise have been restored to Health and Comfort, and become useful to themselves, their Families, and the Publick, for many Years after; and considering moreover, that even the poor Inhabitants of this City, tho' they had Homes, yet were therein but badly accommodated in Sickness, and could not be so well and so easily taken Care of in their separate Habitations, as they might be in one convenient House, under one Inspection, and in the Hands of skilful Practitioners; and several of the Inhabitants of the Province, who unhappily became disorder'd in their Senses, wander'd about, to the Terror of their Neighbours, there being no Place (except the House of Correction) in which they might be confined, and subjected to proper Management for their Recovery, and that House was by no Means fitted for such Purposes; did charitably consult together, and confer with their Friends and Acquaintances, on the best Means of relieving the Distressed, under those Circumstances; and an Infirmary, or Hospital, in the Manner of several lately established in *Great-Britain,* being proposed, was so generally approved, that there was Reason to expect a considerable Subscription from the Inhabitants of this City, towards the Support of such a Hospital; but the Expense of erecting a Building sufficiently large and commodious for the Purpose, it was thought would be too heavy, unless the Subscription could be made general through the Province, and some Assistance could be obtained from the Assembly . . .

Even the title reads with a commanding beat: *Some Account of the Pennsylvania Hospital.*

Benjamin Franklin's activities in promoting institutions for the betterment of the daily lives of his fellow Americans are well known. Readers are aware of his initiative in proposing or supporting America's first circulating or subscription library, the school or academy that has grown into the University of Pennsylvania, a fire company, an insurance company, and a regular constabulary, and the introducing of paved, swept, and well-lit streets in Philadelphia.[4]

All too often, Franklin is presented as a doer of good for the community without indication of the influences in both Boston and Philadelphia that helped to condition his outlook on humanity and its needs. In his autobiography Franklin told of two books "which perhaps gave me a turn of thinking that had an influence on some of the principal future events of my life." These were "a book of Defoe's, called an Essay on Projects, and another of Dr. Mather's, called Essays to do Good."[5] Writing from France at the age of seventy-eight, he expressed his indebtedness to Cotton Mather. In a letter to Cotton's son Samuel,[6] Franklin wrote that as a boy, he had found a copy of Mather's *Essays to Do Good.*[7] "It had been so little regarded by a former possessor, that several leaves of it were torn out; but the remainder gave me such a turn of thinking, as to have an influence on my conduct through life; for I have always set a greater value on the character of a *doer of good,* than on any other kind of reputation; and if I have been, as you seem to think, a useful citizen, the public owes the advantage of it to that book."

Those who think of Cotton Mather as the typical Puritan bigot may hold it odd that he had any influence on the liberal Franklin. Cotton Mather, however, has been a neglected figure in our history, and his full stature and character are not generally known. One revaluation of Mather has come from the study[8] of his interest in science.[9] It has been shown that he was the author of the first account of spontaneous hybridization in plants.[10] His sponsorship of inoculation against the smallpox in 1721 has become part of the literature on colonial American medicine,[11] and his manuscript treatise on medicine, *The Angel of Bethesda,* has been subject to scholarly analysis.[12] Franklin's brother James carried on a running opposition to the Mathers in his *New England Courant* while Benjamin was still an apprentice, and the Franklins marshaled all possible evidence and argument, largely supplied by Dr. William Douglass, against inoculation as part of the anti-Mather campaign. But in later years Franklin became one of the chief advocates of inoculation in colonial America.[13]

Franklin quickly forgot the old hostility to the Mathers once he had moved to Philadelphia. In 1724, on his first return trip to Boston, Franklin had a friendly visit with Cotton Mather, who received him in his library. On the way out, Mather turned suddenly to Franklin and said, "*Stoop, stoop!*" Franklin "did not understand him, till I felt my head hit against the beam [which crossed the narrow passage]. He was a man that never missed any occasion of giving instruction, and upon this he said to me, '*You are young, and have the world before you; stoop as you go through it, and you will miss many hard thumps.*' This advice, thus beat into my head, has frequently been of use to me; and I often think of it, when I see pride mortified, and misfortunes brought upon people by their carrying their heads too high."

What could Franklin possibly have learned from Cotton Mather? At least a love of science. Loving science implied, in Mather's terms, a respect for nature, a recognition that the empirical evidence of the operations of nature in the external world is in perfect harmony with the principles of revelation and true faith. "Be sure," wrote Mather in his instructions to young ministers, "the Experimental Philosophy is that, in which alone your mind can be at all established."[14] The guide in this "experimental philosophy" was "the incomparable Sir Isaac Newton," acknowledged by Mather to be "the Perpetual Dictator of the Learned World in the Principles of Natural Philosophy."[15]

Franklin, of course, had little sympathy with Cotton Mather's theological outpourings. Yet in Mather's activities in "that realm of social experience" called by Max Weber "the Protestant ethic," we may catch a glimpse of Franklin in the making.[16] Mather told his congregation to attend to "some settled business, wherein a Christian should for the most part spend most of his time and this, that so he may glorify God by doing of *Good* for *others,* and getting of *Good* for himself." Puritans in the Boston of Franklin's youth agreed that all the affairs of man are governed by divine providence; thus the gaining of an estate is the result of divine favor. Initiative and industry are, of course, the necessary conditions for accumulating the world's goods, but they are not sufficient. In 1748, Franklin echoed the sentiments of his Puritan masters in his "Advice to a Young Tradesman." The "way to wealth," he wrote, "depends chiefly on two words, *industry* and *frugality,*" but industry and frugality will be of avail only "if that Being who governs the world, to whom all should look for a blessing on their honest endeavours, doth not, in His wise providence, otherwise determine."[17]

"Honor the Lord with thy substance," Cotton Mather preached; "so shall thy barns be filled with plenty." Charitable men are esteemed by God, who rewards them "with remarkable success in their affairs, and increase of their property." Virtue, charity, and hard work in a calling procure divine favor and, therefore, an accumulation of the world's goods. But the man with an estate must remember that he is a steward and he must, therefore, as Cotton Mather put it, glorify God by contributing "unto the welfare of mankind, and such a relief of their miseries as may give the children of men better opportunity to glorify Him."

Mather's *Essays to Do Good* contained the description of "the ravishing satisfaction" which a man "might find in relieving the distresses of a poor, mean, miserable neighbour." Readers were exhorted to do "service for the kingdom of our great Saviour in the world; or any thing to redress the miseries under which mankind is generally languishing." Poor Richard later put it in this way: "Serving God is doing good to man, but praying is thought an easier service, and therefore more generally chosen." On another occasion, Poor Richard said, "The noblest question in the world is, what good may I do in it?" This was merely an improved way of saying what Mather had written in the *Essays:* "Assume and assert the liberty of now and then thinking on the noblest question in the world: What good may I do in the world?"

Franklin's Junto, a private society for the mutual improvement of the members, and a focal point for the introduction of many useful institutions and reforms in Philadelphia, also showed Mather's influence. James Parton was one of the first historians who appreciated that Mather's "Neighborhood Benefit Societies" were the prototype of the Junto and that Mather's "Points of Consideration" (to be read at every meeting) are remarkably similar to the questions asked at the meetings of Franklin's Junto.[18]

Philadelphia, when Franklin arrived there in 1723, was characterized by a spirit of liberal tolerance that seems in marked contrast to theocratic Boston. But however far apart Calvinism and Quakerism might be on doctrinal or theological issues, they shared in common the Protestant ethic. Franklin, therefore, probably found in Philadelphia a familiar outlook on social goals and the rewards of virtuous conduct. Puritan Boston, in other words, prepared him for his life in Philadelphia. Philadelphia and Benjamin Franklin made an ideal combination and both grew together—Franklin to become the foremost citizen of the New World, and Philadelphia to become the third largest city (not including those of India) in the eighteenth-century British Empire.[19] So similar were the precepts common in Quaker Philadelphia and Puritan Boston that both were put into the mouth of Poor Richard and what came out sounded just like Benjamin Franklin. "Diligence is a 'virtue' useful and laudable among men . . . First, it is the Way to Wealth . . . Frugality is a virtue too, and not of little use in life, the better Way to be Rich, for it has less toil and temptation. It is proverbial, *A penny saved is a penny got.*"[20] These lines, excerpted from an essay on economic virtues written by William Penn, might have been written by Cotton Mather, and were actually rewritten by Franklin for Poor Richard. Philadelphia Quakers, like their Puritan brethren in Boston, sought wealth and position for "the Honour of God and Good of Mankind." They too considered that wealth conferred an obligation of stewardship upon the man with an estate. As William Penn expounded the doctrine, "But of all we call ours, we are most accountable to God and the publick for our estates: In this we are but stewards, and to hoard up all to ourselves is great injustice as well as ingratitude."

Most discussions of Franklin indicate that his career and social outlook are to be considered in sharp differentiation from the prevailing temper of Boston and New England generally, because Franklin's view of society was secular and not theological. It is certainly true that Franklin absorbed the Protestant ethic without maintaining any ties with the Calvinist theology in which it had been imbedded in the Boston of his youth. Franklin may, therefore, be considered to have demonstrated that the Protestant ethic itself could survive without that particular theology. Cotton Mather was not the only Boston preacher to have held ideas which were similar to Franklin's but the fact remains that Franklin actually knew Mather, read his works, and more than once insisted on Mather's great influence upon him. Throughout all of Franklin's writings there appear phrases which maintain the spirit of the preachings of Mather and his contemporaries and which are testimony to their lasting effect upon him.

Franklin's program of good works and public improvements evoked a congenial and sympathetic response from the public-spirited citizens of Philadelphia. There was a regular plan of action. First there would be a paper read by Franklin at a meeting of the Junto, where it would be discussed; its contents would then be communicated to the associated Juntos; and, finally, one or more editorials would appear in Franklin's *Pennsylvania Gazette.* Once the public's attention had been called to the problem, some action might be expected, especially under the prodding of the newspaper and the network of Junto organizations. In this way, for example, Franklin succeeded in establishing a regular constabulary. He tells us how he "wrote a paper to be read in the Junto" in which he proposed "the hiring of men to serve constantly in that business," their salaries to be paid by a property tax. "This idea, being approved by the Junto, was communicated to the other clubs, but as [if] arising in each of them; and though the plan was not immediately carried into execution, yet, by preparing the minds of people for the change, it paved the way for the law obtained a few years after, when the members of our clubs were grown into more influence."[21] The Union Fire Company and the Academy were introduced in this way and so were other public improvements.

Franklin soon became a master of the art of promotion, and everyone with a project sought out his aid. When the Reverend Gilbert Tennent wished to obtain subscriptions for building a "meeting-house . . . for the use of a congregation he had gathered among the Presbyterians," he came to Franklin for assistance. But Franklin was "unwilling to make myself disagreeable to my fellow-citizens by too frequently soliciting their contributions," he explained to Tennent, though he was willing to give advice. "I advise you," he told Tennent, "to apply to all those whom you know will give something; next, to those whom you are uncertain whether they will give anything or not, and show them the list of those who have given; and, lastly, do not neglect those who you are sure will give nothing, for in some of them you may be mistaken." The advice was sound, Franklin was proud to observe, and he pointed in evidence to "the capacious and very elegant meeting-house that stands in Arch-street."[22]

In his autobiography, Franklin related the history of the founding of the Pennsylvania Hospital as follows:

> In 1751, Dr. Thomas Bond, a particular friend of mine, conceived the idea of establishing a hospital in Philadelphia (a very beneficient design, which has been ascribed to me, but was originally his), for the reception and cure of poor sick persons, whether inhabitants of the province or strangers. He was zealous and active in endeavouring to procure subscriptions for it, but the proposal being a novelty in America, and at first not well understood, he met with but small success.
>
> At length he came to me with the compliment that he found there was no such thing as carrying a public-spirited project through without my being concerned in it. "For," says he, "I am often asked by those to whom I propose subscribing, Have you consulted Franklin upon this business? And what does he think of it? And when I tell them that I have not (supposing it rather out of your line), they do not subscribe, but say they will consider of it." I enquired into the nature and probable utility of his scheme, and receiving from him a very satisfactory explanation, I not only subscribed to it myself, but engaged heartily in the design of procuring subscriptions from others. Previously, however, to the solicitation, I endeavoured to prepare the minds of the people by writing on the subject in the newspapers, which was my usual custom in such cases, but which he had omitted.
>
> The subscriptions afterwards were more free and generous; but, beginning to flag, I saw they would be insufficient without some assistance from the Assembly, and therefore proposed to petition for it, which was done. The country members did not at first relish the project; they objected that it could only be serviceable to the city, and therefore the citizens alone should be at the expense of it; and they doubted whether the citizens themselves generally approved of it. My allegation on the contrary, that it met with such approbation as to leave no doubt of our being able to raise two thousand pounds by voluntary donations, they considered as a most extravagant supposition, and utterly impossible.
>
> On this I formed my plan; and, asking leave to bring in a bill for incorporating the contributors according to the prayer of their petition, and granting them a blank sum of money, which leave was obtained chiefly on the consideration that the House could throw the bill out if they did not like it, I drew it so as to make the important clause a conditional one, viz., "And be it enacted, by the authority aforesaid, that when the said contributors shall have met and chosen their managers and treasurer, *and shall have raised by their contributions a capital stock of*   *value* (the yearly interest of which is to be applied to the accommodating of the sick poor in the said hospital, free of charge for diet, attendance, advice, and medicines), *and*

> *shall make the same appear to the satisfaction of the speaker of the Assembly for the time being,* that *then* it shall and may be lawful for the said speaker, and he is hereby required, to sign an order on the provincial treasurer for the payment of two thousand pounds, in two yearly payments, to the treasurer of the said hospital, to be applied to the founding, building, and finishing of the same."
>
> This condition carried the bill through; for the members, who had opposed the grant, and now conceived they might have the credit of being charitable without the expense, agreed to its passage; and then, in soliciting subscriptions among the people, we urged the conditional promise of the law as an additional motive to give, since every man's donation would be doubled; thus the clause worked both ways. The subscriptions accordingly soon exceeded the requisite sum, and we claimed and received the public gift, which enabled us to carry the design into execution. A convenient and handsome building was soon erected; the institution has by constant experience been found useful, and flourishes to this day; and I do not remember any of my political manoeuvres, the success of which gave me at the time more pleasure, or wherein, after thinking of it, I more easily excused myself for having made some use of cunning.[23]

Franklin put into practice the very advice he gave to the Reverend Tennent; how well it succeeded!

In his book about the hospital, Franklin reprinted the newspaper articles about the needs for a hospital. They are not to be found in Franklin's works. The portion of the bill he drew up for the Assembly, quoted in the above extract from the autobiography, does not agree exactly with the text as printed by Franklin in his book. Franklin does not even mention the book in his autobiography.

Carl Van Doren described the style of Franklin's *Some Account of the Pennsylvania Hospital* in these words: "His opening sentence is an example of homespun splendor hardly to be matched in the English language. The great sentences with which writers begin books commonly make use of flaming words; Franklin's words are all plain. His magic comes from his cadence and the emotion it implies."[24] Each reader may verify Van Doren's comments by reading the opening paragraphs of Franklin's book.

The history of the Pennsylvania Hospital has been rewritten many times since Franklin published his account. In 1951, on the occasion of the bicentenary of the founding, a number of historical articles were published.[25] In 1953, Edward B. Krumbhaar contributed a brief history to a volume of essays on "historic Philadelphia."[26] In 1938, Francis R. Packard published a charming volume, written in Franklin's fashion and printed in a style of typography and format resembling Franklin's book, entitled *Some Account of the Pennsylvania Hospital from its first Rise to the Beginning of the Year 1938*,[27] thus supplementing and bringing up to date Thomas G. Morton's *History of the Pennsylvania Hospital, 1751–1895*.[28] A continuing stream of publications makes it unnecessary to recount here the story of the hospital from Franklin's day to ours. Still ministering to the needs of the community, but now providing many services not even dreamed of two hundred years ago, the Pennsylvania Hospital stands as a monument to the collaboration of Franklin and Bond and the warm response they were able to evoke from their fellow citizens.

The original cornerstone, with text written by Franklin, can still be read plainly by visitors to the Pennsylvania Hospital:

IN THE YEAR OF CHRIST
MDCCLV.
GEORGE THE SECOND HAPPILY REIGNING
(FOR HE SOUGHT THE HAPPINESS OF HIS PEOPLE)
PHILADELPHIA FLOURISHING
(FOR ITS INHABITANTS WERE PUBLICK SPIRITED)
THIS BUILDING
BY THE BOUNTY OF THE GOVERNMENT,
AND OF MANY PRIVATE PERSONS
WAS PIOUSLY FOUNDED
FOR THE RELIEF OF THE SICK AND MISERABLE:
MAY THE GOD OF MERCIES
BLESS THE UNDERTAKING.

The second history of the Pennsylvania Hospital was a part of the campaign to raise funds in 1759 and the years immediately following. The needs were so much greater than the available funds that the situation was desperate; a petition of the Assembly in 1760 was unsuccessful. (Certainly the Managers missed the aid of Franklin, who was abroad in England on province business.) An appeal was sent to the *Pennsylvania Gazette* and prints or pictures were issued. Finally, a committee was appointed to write a book like Franklin's; the chairman was Samuel Rhoads. Called *Continuation of the Account of the Pennsylvania Hospital; from the first of May, 1754, to the fifth of May, 1761*, the new book described the activities of the hospital, the care given to patients, and the sources of support and income along with the expenses.[29] Whether the book was the effective instrument in obtaining the needed funds, we do not know;[30] but the Assembly members visited the hospital, were favorably impressed by what they saw, and granted the sum of three thousand pounds to get the hospital out of its financial difficulties. Both Franklin's original publication and the *Continuation* were reprinted in 1817 at Philadelphia, "Printed at the Office of the United States' Gazette," and this edition seems to be even rarer than the original.[31]

No one can turn the pages of Franklin's book without being struck by the modernity which characterizes every page, particularly the devices he employed to raise funds. Our admiration is aroused by the account of the noble physicians who served without compensation. The arguments advanced by Franklin in support of the hospital are compelling, and it is difficult to believe that they were conceived two hundred years ago, and not yesterday. Here is presented, in the strongest terms, Franklin's thesis that an organized group of men working for a common goal can do more good in the world than if each man worked to help his neighbors by himself. As an empiricist, Franklin was pleased to be able to demonstrate—by facts and figures—that the community as a whole would receive benefit from the new institution. His argument, that the function of a hospital is to restore sick individuals to their useful place in society, has a twentieth-century ring to it and seems especially appropriate to a free and democratic way of life. On the practical side, there is his suggestion that one of the major services rendered to the community by a hospital is to provide a training ground for young and inexperienced physicians. Physicians from the whole province of Pennsylvania would become acquainted with a greater variety of diseases than they might otherwise encounter from day to day, Franklin argued, and so the hospital would tend to raise the standards of medicine throughout the whole of Pennsylvania. As a staunch republican, Franklin was proud to observe that the hospital "gave to the beggar in America a degree of comfort and chance for recovery equal to that of a European prince in this palace." The hospital placed the patient in better and cleaner surroundings than a private home, and so was more conducive to the patient's recovery; even so, the cost of hospital care was only one-tenth of the cost of similar care in "private lodgings."

The title page of *Some Account of the Pennsylvania Hospital* indicates that the book was "printed by B. Franklin and D. Hall." It was characteristic of Franklin's shrewdness in business that he saw the great advantages that would derive from partnerships spread all over British America. In almost every instance, Franklin was a "silent partner"; each shop operated under the name of the local printer. By 1743, Franklin had established three printing houses (in Philadelphia; Charleston, South Carolina; and New York) and was planning a fourth. Later establishments were set up in Antigua; Lancaster, New Jersey; and Connecticut.[32] Hall, who became Franklin's partner in name as well as in fact, had learned the art of printing in Edinburgh and was working in London in

1743 when he was recommended to Franklin. Franklin invited him to come to Philadelphia and promised either to make him partner or to give him a year's employment and a return passage to England should he wish to leave Philadelphia. David Hall proved to be "obliging, discreet, industrious, and honest." He became foreman of the printing shop in Philadelphia and in 1748 took over the major responsibility as partner. For eighteen years, until 1766, the firm was known as Franklin and Hall, and under Hall's management brought Franklin an average income of 467 pounds per annum.[33]

Franklin did not in general create strikingly new typographical designs, although his best printing and that of his disciples has a style that has been described as "characteristic of Franklin," based on a feeling for the "integrity of type."[34] One of his first "excursions from the well-beaten path" was his *Cato Major,* published in 1744, and it may be conjectured that Hall had a hand in its design. On the other hand, most of the books issued under the Franklin and Hall imprint were inferior in workmanship to those earlier works printed by Franklin himself. We do not know how much of a hand Franklin actually had in the design of *Some Account of the Pennsylvania Hospital,* published in 1754, six years after his retirement. We know that in the fall of 1753 Franklin was again intensely interested in the art of printing, and suggested an improvement in printing presses. Perhaps this revived concern for press problems in 1753 arose from a new contact with actual bookmaking, just before the 1754 publication about the Pennsylvania Hospital. We have no way of telling whether David Hall could have designed this beautiful book by himself, or whether the typographical design was Franklin's. Considering the purpose for which the book was intended, and Franklin's strong personal interest in the project, it is hardly likely that Franklin would merely have turned over the copy to Hall without any further concern about the design and the presswork. In any event, this book displays some typical features of Franklin's printing, with a page well-designed but in too small a type for the size of the printed area and the length of line—a disappointment after the "monumental titlepage."[35] Even so, the pages have a certain typographical elegance. From one page to the next there is a variation in the form in which the printed material is presented. An engaging use is made of italics, capital and small capital letters, and running heads or captions. Even mere lists of names are presented in a contrasting manner, and statistical data are printed in an attractive fashion. Despite the limitations that derive from an economy in paper, and admitting that the type size might have been somewhat larger, there is little question but that Franklin's superb sense of craftsmanship made the book as a whole attractive to the eye and pleasant to read.

The institution that this book was intended to serve was never "the lengthened shadow of one man." From the very beginning, the Pennsylvania Hospital thrived on the public-spirited devotion of a group of persons with a common purpose, expressed in the original seal, bearing the device of the Good Samaritan conveying the sick man to an inn and the words "Take care of him, & I will repay thee." Many innovations in American medical teaching and public medicine originated in the Pennsylvania Hospital, among them clinical teaching, the outpatient department, and "one of the first group practice clinics in America organized for comprehensive diagnostic service on a single-fee basis."[36] The Managers of the Pennsylvania Hospital have always considered the public interest first, making services available equally to all, rich or poor, and patients both afoot or abed. Even "home care" was provided, beginning in 1807, from the hospital and in 1808 the policy was adopted that "the poor of Pennsylvania shall be vaccinated gratis, if they will call at the Hospital."[37]

Conceived by Thomas Bond, brought into existence through the aid of Benjamin Franklin, the Pennsylvania Hospital continues to serve the country as the pioneer example that the efforts of private citizens can produce permanent institutions of public benefit. As we turn the pages of the first appeal for public support of the Pennsylvania Hospital, we may recapture the original spirit that motivated the people of Franklin's day to minister to the needs of their neighbors. Poor Richard said in 1738, "If you would not be forgotten as soon as you are dead and rotten, either write things worth reading or do things worth the writing." The Pennsylvania Hospital and Franklin's book about it indicate that he did both.

11

The Transit of Mercury

Franklin's activities for the general promotion of science exhibit not only a concern for increasing knowlege but also his firm conviction that science must be an international pursuit. It was in this spirit that he issued a passport in 1779 asking "all Captains and Commanders of armed Ships acting by Commission from the Congress of the United States of America, now in war with Great Britain," not to consider Captain Cook's ship "as an Enemy, nor suffer any Plunder to be made of the Effects contain'd in her, nor obstruct her immediate return to England, by detaining her or sending her into any other Part of Europe or to America." He urged that they "treat the said Captain Cook and his People with all Civility and Kindness, affording them, as common Friends to Mankind, all the Assistance in your Power, which they may happen to stand in need of." Franklin explained that Captain Cook ("that most celebrated Navigator and discoverer") was a scientist returning from a scientific expedition, one "truly laudable" since "the Increase of Geographical Knowledge facilitates the communication between distant Nations, in the Exchange of Useful Products and Manufactures, and the Extension of Arts, whereby the common Enjoyments of human Life are multiply'd and augmented, and Science of other kinds increased to the benefit of Mankind in general."

Franklin's activities in relation to the transit of Mercury of 1753 show the same recognition that science requires a worldwide effort in which all civilized nations must participate. But it is significant that Franklin expressed his hope that on this occasion "we may not be excelled by the French, either in diligence or accuracy." That is, as he wrote to James Bowdoin, he hoped that "our country will not miss the opportunity of sharing the honor to be got on this occasion." Here is an early expression of the need for Americans to do their share in the international science effort and to gain a proper place in the international scientific community. It was the same kind of national pride that, at least in part, impelled him toward the founding of a scientific or "philosophical" society in Philadelphia.

Information concerning Franklin and the transit of Mercury is drawn chiefly from a pamphlet which I found during the course of my investigation of the scientific career of Benjamin Franklin. It is part of the extensive Franklin material collected by Jared Sparks, at present in the Harvard Library. This pamphlet, relating to the transit of Mercury that occurred in 1753, occupies four printed pages and is supplemented by a hand-drawn diagram.

For about a century, the only source of information concerning this pamphlet was a footnote written by Sparks for his edition of *The Works of Benjamin Franklin.*[1] On the basis of the information collected by Sparks, A. H. Smyth, in his edition of *The Writings of Benjamin Franklin,* wrote a short note giving the title of the work.[2] Smyth copied Sparks's transcription of the title as "Letters relating to a Transit of Mercury over the Sun, which is to happen May 6th, 1753." That Smyth actually copied the title from Sparks may be seen from the following facts: Sparks made an error in transcription ("a Transit" rather than "the Transit"), an error not likely to have appeared also in Smyth had the latter taken his title from the original; Smyth quoted the title in italics, as well as enclosing it within quotation marks, just as it appeared in Sparks, although this was not Smyth's custom in other citations; and Smyth declared that the letter to which his note was appended was "First published by Sparks." Smyth's note reads:

> The paper alluded to, of which fifty copies were struck off for distribution, was entitled, "*Letters relating to a Transit of Mercury over the Sun, which is to happen May 6th, 1753,*"—Ed.

In Campbell's "Short-Title Check List," the title is given as in Smyth, with a reference to "Smyth's Franklin, vol. 3, p. 122. Not in Hildeburn. tc." The letters "tc" signify that the "full title or collation is unknown."[3] As Campbell's note indicates, this publication is not mentioned in Hildeburn's *Issues of the Press of Pennsylvania.* Curiously enough, had Campbell consulted Sparks for his information rather than Smyth, he would have learned that the collation is as follows: "It consisted of four large folio pages . . . A manuscript drawing was also attached to each copy, showing the line in which Mercury would pass over the sun's disc." This example well illustrates the need for consulting Sparks as well as Smyth whenever investigating a question concerning Franklin, for not only are there many documents contained in Sparks not to be found in Smyth, but there are also many interesting and important notes written by Sparks that are of inestimable value. The magnificent new edition of Franklin, sponsored jointly by the American Philosophical Society and by Yale University similarly does not render the earlier editions (notably those of Sparks and Smyth) wholly redundant.

Yet another reference (in print) to this pamphlet occurs in Charles Evans, *American Bibliography.*[4] The title is given correctly although Evans did not know that Franklin was the printer. A single copy is located at the Massachusetts Historical Society. However, following the title, one finds "[Boston: 1753]" signifying that although the place of publication was not stated explicitly on the pamphlet, it was presumed to have been published in Boston; actually, it was printed in Philadelphia, and by Benjamin Franklin. Next, Evans gives the pagination as "pp. [4]" which, in the style of his catalogue, may mean 4 unnumbered pages, whereas this pamphlet is actually paginated "[1]"; "[2]"; "[3]"; "[4]"; followed by a page containing a manuscript drawing. In Worthington C. Ford's *Broadsides, Ballads, etc. Printed in Massachusetts 1639–1800,* this pamphlet is attributed to John Winthrop, Hollis Professor of Mathematics and Natural Philosophy at Harvard College; although Ford does not give a place of publication, the inclusion of this pamphlet in the volume indicates that it is supposedly a Massachusetts imprint. Finally, in Joseph Sabin's *Bibliotheca Americana,* the pamphlet is also attributed to Winthrop and the place of publication is given as Boston. Thus, this pamphlet printed by Franklin in Philadelphia has been often described, but not correctly.[5]

Franklin described the pamphlet in a letter to James Bowdoin written from Philadelphia, 28 February 1753:

> The enclosed is a copy of a letter and some papers I received lately from a friend, of which I have struck off fifty copies by the press, to distribute among my ingenious acquaintances in North America, hoping some of them will make the observations proposed. The improvement of geography and astronomy is the common concern of all polite nations, and, I trust, our country will not miss the opportunity of sharing in the honor to be got on this occasion. The French originals are despatched by express overland to Quebec. I doubt not but that you will do what may lie in your power, to promote the making [of] these observations in New England, and that we may not be excelled by the American French, either in diligence or accuracy. We have here a three-foot reflecting telescope, and other proper instruments; and intend to observe at our Academy, if the weather permit. You will see, by our Almanac, that we have had this transit under consideration before the arrival of these French letters.[6]

Franklin's letter to Bowdoin mentions that the "French originals are despatched by express overland to Quebec." According to Sparks's note, a translation of the French documents

> was made in New York under the direction of Mr. James Alexander, who sent them to Franklin. The French astronomers were desirous, that observations of the transit should be taken at Quebec. M. de Lisle, of the Academy of Sciences, drew up a memorial containing instructions for the purpose. This memorial, with letters from M. La Gallissonière, dated at Paris, October 10th, 1752, was sent unsealed to the governor of New York, with a request that they might be forwarded over land to Quebec. The governor put the papers into the hands of Mr. Alexander, who caused a translation to be made. The originals were then dispatched to Quebec. To the translation, which was printed and distributed by Franklin, was prefixed a long and interesting letter from Mr. Alexander on the subject of the transit. He also communicated a copy of all the papers to Cadwallader Colden.

The concluding sentence of the paragraph in Franklin's letter to Bowdoin refers to "Poor Richard's Almanack" for 1753. Under the heading "Eclipses, 1753," Franklin described the coming phenomenon:

> On *Sunday*, the 6th Day of *May*, in the Morning, the Planet *Mercury* may be seen to make a black Spot in the *Sun's* Body, according to the following Calculation [omitted].
>
> The astronomical Time when *Mercury* goes off the *Sun's* Disk, being reduced to common Time, is *May* the 6th, at 31 min. after Seven in the Morning. The *Sun* rises at 1 min. past Five, and if you get up betimes, and put on your Spectacles, you will see *Mercury* rise in the *Sun*, and will appear like a small black Patch in a Lady's Face.[7]

This is followed by a diagram showing the path of Mercury as it crosses the sun's disc.

The pamphlet printed by Franklin on the occasion of the transit is made up as follows: pages 1 and 2 contain the letter from James Alexander to Franklin of 26 January 1753, printed without date and without Alexander's name; at the bottom of page 2 and continuing through the top of page 3 are "TRANSLATIONS *of the Letters,* &c. *from* Paris," first the letter from La Galissonnière, dated Paris 10 October 1752, printed with the date, but without La Galissonnière's name, followed by La Galissonnière's letter of the same date *"To the Reverend, Reverend Father,* Bone Camp, *Jesuit, at* Quebeck," printed without La Galissonnière's name; the remainder of page 3 and all of page 4 are devoted to the "MEMORIAL, *mentioned in the last. The Passage of* MERCURY *upon the* SUN, *on the 6th of* May, 1753, *to observe in* Canada." It will be noted that Franklin's title for the whole pamphlet refers to the "transit of Mercury over the Sun," whereas Alexander's translation of the "memorial" is a literal rendering, "passage of Mercury upon the Sun." (The reason for the suppression of the names of the correspondents will presently be made evident from Alexander's correspondence.)

The memorial concerning the transit had been drawn up by Joseph-Nicolas Delisle (or De l'Isle), astronomer, naval geographer, professor of mathematics in the Collège Royal de France, and had been sent by Roland-Michel Barrin, marquis de la Galissonnière, whose distinguished career embraced the posts of lieutenant general of the navy and governor general of Canada.[8]

A letter from Alexander to Franklin, dated New York, 29 January 1753, makes clearer the circumstances under which Alexander made his translation. It also explains that the omission of La Gallissonnière's name was a condition of the "leave" given to print the documents.

> About a week ago the papers whereof that herewith Mark't A is a Translation were sent to Peruse, and on my returning them With a Translation I beged Leave to Communicate Coppies of them to You and Dr. Colden with Liberty if you thought proper to print them or any Part of them. In Answer To which I obtained the Leave coppied at the end of Them, and Struck out La Galissonnière's name, for the reason in the Leave.
>
> The Delay of that Leave put me upon dipping into the Abridgment of the Philosophical Transactions to see what there said of the matters in those papers, and from thence I made the Extracts in the paper Herewith Mark't B, by which I find that that which had baffled all the Art of man hitherto To discover with any Toleable Certainty (Viz: the Suns Distance from the Earth) may with great Certainty be Discovered by the Transit of Venus over the sun the 26 of May 1761 old stile if well Observed in the East Indies and here and there Observations Compared Together.
>
> It Would be a great honour To our young Colledges in America if they forthwith prepared themselves with a proper apparatus for that Observation and made it. Which I Doubt not they would Severally Do if they were Severally put in mind of it and of the great Importance that that Observation would be To Astronomy and that the missing that One Observation cannot be retrieved for 250 years To come.
>
> You have on so many Occasions Demonstrated Your Love to Literature and the good of Mankind in General that I thought no person so proper as your self to think of the ways and means of persuading these Colledges to prepare themselves for takeing that Observation and in order to it you may make what use you please of the papers herewith, only not my name.[9]

It is clear from this letter that Alexander thought of having observations made of the transit of Mercury chiefly to arouse interest in the transit of Venus of 1761 and to afford practice for that latter observation. The importance of an observation of the transit of Venus in the middle of the eighteenth century can hardly be exaggerated. Newtonian mechanics, founded on Kepler's laws of planetary motion had been expressed in terms of the relative distances of the planets from the sun. In other words, discussions of the planets Mercury, Venus, Mars, Jupiter, and Saturn never referred to the actual distances of these planets from the sun, say in miles or feet, but rather to their distances from the sun compared to the earth's distance from the sun. Thus, if the earth's distance from the sun were taken arbitrarily at 1,000 units, then Mercury's distance from the sun was reckoned at 387, and Venus's at 723.[10] Hence, if it were possible to determine the solar parallax—the angle subtended at the sun by the earth's radius—one could compute first the actual distance from the earth to the sun, and then the actual distances of the several planets from the sun. A method for determining the solar parallax, usually attributed to Edmond Halley, used observations, made at different places on the earth, of the time in which Venus was seen to cross the face of the sun. This phenomenon, a transit of Venus across the sun, is an infrequent occurrence. Since 1600, transits of Venus have occurred in 1631, 1639, 1761, 1769, 1874, and 1882. The next pair will be in 2004 and 2012. The transit of 1631 occurred during the night, and that of 1639 was imperfectly observed. Hence in 1761, there occurred one of a pair of events, to be repeated in eight years, and then not again for some 120 years more, of supreme importance for astronomical science.

Franklin himself was well aware of the importance of the transit of Venus, even before receiving Alexander's letter. In *Poor Richard* for 1753, the reader was told that the moon is sufficiently close to the earth to yield "a very sensible Parallax" and thus astronomers "know certainly the distance of the Moon from the Earth, *viz.* 240 thousand Miles." After declaring that the sun's distance from the earth is "at least . . . eighty Millions of Miles," *Poor Richard* continues, "but it is not certainly known, whether it is not a great deal more. In the Year 1761, the Distance of all the Planets from the Sun will be determined to a great Degree of Exactness by Observations on a Transit of the Planet *Venus* over the Face of the Sun, which is to happen the 6th of *May,* O.S. in the Year."[11] When the transit of Venus actually occurred, the most notable American feature of that event was the Harvard professor John Winthrop's voyage to Newfoundland aboard the Province sloop, furnished for that purpose by Massachusetts's Governor Bernard—the first scientific expedition to be sponsored by a college in America—which provided the only American observation of the transit of 1761.[12] Mercury is too close to the sun to yield the solar parallax; the transit of Mercury merely served as a check on planetary computations and as a means of checking longitude determinations.

Alexander sent copies of the documents relating to the transit to his friend Cadwallader Colden as well as to Franklin, expressing the hope that Colden "could be here Some days before the 6th of May next to assist in prepareing things for the observing the transit of Mercury over the Sun then, and in makeing the observation, for Except your Self & me, I believe there's none in the province any way acquainted with observations of that kind, and our observing that transit might show some young men how to observe the transit of Venus in 1761."[13]

Franklin, meanwhile, was preparing to observe the transit in Philadelphia. On 28 February 1753, he told Colden that "We are preparing here to make accurate Observations on the approaching Transit of Mercury over the Sun," and asked for certain information.[14] On 12 April 1753, he wrote to Colden thanking him "for the Hints concerning our Observation of the Transit. I wish we may have fair Weather for I think nothing else will now be wanting."[15]

The English scientists were anxious to receive observations from the New World. Peter Collinson, friend and patron of Franklin, John Bartram, and Colden, wrote to the latter from London on 2 June 1753 that "Wee hope to be favour'd with your observation on the Transit of Mercury."[16] On 1 September 1753, Collinson wrote again to Colden, saying, "I have no Letter Since yours of February 16 pray did you make any observations on the Transit of Mercury"?[17]

Alexander received the pamphlet printed by Franklin in March 1753, and wrote to Colden on 4 March:

> Yesternight I received by the post, some printed Coppies of the Letters concerning the transit of Mercury one of which I Send you herewith . . .
>
> Mr. Franklin writes that he has struck off fifty coppies of the Letters, & by this post Sends same to Jersey & New England & by next Southern post will Send some to Maryland & Virginia all with pressing Letters from himself to provide for observing the transits mentioned within—I hope your affairs may permitt you to give us your assistance here on May 6th next.[18]

Alas! Alexander's preparations came to nought and he wrote, sadly, to Colden on 10 May 1753, "I had this day the favour of yours of the 6th the clouds debarred us here of the Light of the transit of Mercury though we were pretty well prepared for the observation."[19]

Franklin himself had no better luck. I have been unable to find Franklin's account of his own attempt to observe the transit. Yet Peter Collinson heard about it and also about Franklin's endeavors to obtain observations from others. In a letter dated 12 August 1753 (or 1754), Collinson wrote to Franklin that he was disappointed at the bad luck that attended the transit of Mercury, adding that Franklin's zeal to promote that observation could not be enough commended.[20] Furthermore, in a letter from Alexander to Colden, dated 30 July 1753, we find "I rec'd by this post from Mr. Franklin who is now at Boston a Coppy of an observation of the transit of Mercury at Antigua which I believe is the only one in the British Colonies—of which inclosed is a Coppy—I found in the Boston Evening post of July 23d an observation of the same at London of which inclosed is also a Coppy."[21]

The single observation referred to by Alexander was made in Antigua by William Shervington (or Shewington) or perhaps by Captain Richard Tyrel.[22] The letter which follows, apparently representing the only observation of the transit made in British America, is reprinted from the *Philosophical Transactions,* the official journal of the Royal Society of London. The two notes are included in the *Philosophical Transactions.*[23]

A Letter of Mr. William Shervington to Benjamin Franklin, Esq; of Philadelphia, concerning the Transit of Mercury over the Sun, on the 6 of May 1753, as observed in the Island of Antigua: Communicated by Mr. Peter Collinson, F.R.S.

Antigua, June 20, 1753; Read Nov. 15, 1753

Sir,

Mr. Benjamin Mecom having received half a dozen circulatory letters from you relating to Mercury's transit over the sun the 6 of last May, he put them into my hands. One would have sufficed for our island, as we are not overburthen'd with men, who have a taste that way. Hereunder I send you the result of my observation thereof.

Sunday, May 6th, at $6^{h}\ 7'\ 51''$, I observed the western limb of Mercury to touch the western limb of the sun; and at $6^{h}\ 10'\ 37''$, he touch'd the same with his eastern limb, and totally disappear'd. Lat. of the place $17°\ 0'$ N. Lon. by estimation $61°\ 45'$ W. from London.

This was taken by a Graham's watch, and corrected by two altitudes taken by a most exquisite quadrant; *viz.*

At $6^{h}\ 58'\ 7''$ I observed the distance of the sun's upper limb from the zenith $= 72°\ 21'\ 30''$. And at $9^{h}\ 31'\ 5''$, I observed the same $= 36°\ 17'\ 0''$.

By the common process (which you may have, if necessary) I found the watch was $0°\ 4'\ 4''\ 28'''$ too fast, therefore,

	From $6^{h}\ 10'\ 37''$
	Take $0^{h}\ 4'\ 4''\ 28'''$
True apparent time of Mercs. exit here,	$6^{h}\ 6'\ 32''\ 32'''$

Pray impart your observation to

Your Well-Wisher,
William Shervington.[24]

Clearly the total effect of Franklin's activity in printing and distributing the pamphlet on the transit of Mercury did not greatly advance our knowledge of the universe. We see, however, in this story yet another example of Franklin's general interest in all branches of science and his zeal in advancing the cause of science by his personal efforts whenever the occasion offered itself. Franklin's double role in the development of science on the American continent consisted first, of adding to knowledge by his own penetrating research and second, of stimulating others to do research, organizing scattered individual scientific efforts so that they might become more effective, and transmitting scientific information to his countrymen and his fellow scientists abroad.[25]

12

Faraday and the "Newborn Baby"

Scientists who have been questioned about the utility of their work often respond with some variant on the reply of Benjamin Franklin on such an occasion. What good, he is said to have asked, is a newborn baby? In Seymour Chapin's admirable study of Franklin's bon mot, a mention is made of the attribution to Michael Faraday of the phrase about the newborn baby.[1] Faraday is often supposed to have used some variant of the question "What good is a newborn baby?" in relation to his own discoveries in electromagnetism.[2] In fact, Faraday did use this metaphor, but in a quite different context.

In 1816, when Faraday made use of Franklin's apothegm, he had in mind the recent advances in chemistry. The occasion was a lecture before the City Philosophical Society of London, "On Oxygen, Chlorine, Iodine and Fluorine."[3] The lecture dealt in large measure with the important halogen family of chemical elements, recently discovered or identified as elements by Humphry Davy. Faraday's text reads, in part: "Before leaving this substance, chlorine, I will point out its history, as an answer to those who are in the habit of saying to every new fact, 'What is its use?' Dr. Franklin says to such, 'What is the use of an infant?' "[4] Here Franklin's question is asked in a slightly different form than is common.

Faraday not only asked the Franklinian question, in a rhetorical fashion; he also answered it directly. "The answer of the experimentalist," he continued, would be " 'Endeavour to make it useful.' When Scheele discovered this substance it appeared to have no use, it was in its infantine and useless state; but having grown up to maturity, witness its powers, and see what endeavours to make it useful have done."

Humphry Davy announced in 1810 that the substance obtained some decades earlier by the Swedish chemist Scheele must be an element.[5] Davy gave it its name from its greenish color. The use to which Faraday referred was not specified by him. We know of two quite different uses to which the discovery was put. One was theoretical. The then-current system of chemistry, formulated primarily by Lavoisier, contained a postulate to the effect that all acids contain oxygen. Because of its special role as acid-producer this gas, discovered by Scheele and by Priestley, had been given the name oxygen—from Greek roots represented by the adjective *oxys* (sharp, pungent, acid) and the suffix *genes* (born, produced). Circulated in the *Nomenclature chimique* (Paris, 1787) by Lavoisier, Guyton de Morveau, and their associates, *oxygène* thus signified the acidifying principle.[6]

Common muriatic acid had been decomposed into hydrogen and the substance we now, following Davy, call chlorine. This substance was then thought to be—nay, had to be—a compound of oxygen and something else, if Lavoisier's postulate was valid. But when Davy showed that muriatic acid is composed solely of the chemical elements hydrogen and chlorine, a central part of Lavoisier's theory had to be discarded.[7]

A second use of chlorine was in the textile industry. Various compounds of chlorine had been in use as bleaching agents, especially after the introduction in 1798 of a method of manufacturing bleaching powder, invented by Charles T. Tennant of Glasgow.[8] This may have been the use Faraday had in mind, since he refers to Scheele's discovery of "this substance" rather than Davy's identification of it as a chemical element. Faraday himself, working to a degree under Davy's direction, had liquefied chlorine within a year of the latter's announcement that this substance is an element.[9]

There is a second anecdote associated with Faraday and the possible use of a new discovery. In this one, the Prime Minister is said to have asked Faraday what use his discovery of electromagnetism might have. This time, his reply was supposed to have been: "Soon you will be able to tax it." Unlike for the first anecdote, I have been unable to find any direct authentication of this second one. On the face of it, the story seems unlikely. We today, on hearing the term electromagnetism, think at once of electric generators and motors, but at that time the vast practical industry that has been reared upon Faraday's fundamental discoveries was still to be born. He declared in 1831, "I have rather been desirous of discovering new facts and new relations dependent on magneto-electric induction, than of exalting the force of those already obtained; being assured that the latter would find their full development hereafter."[10]

This story, however, may be traced back to the historian William Edward Hartpole Lecky. In the introduction to his *Democracy and Liberty,* Lecky discussed Gladstone's attitude toward science and scientific discoveries:

> There were, it is true, wide tracts of knowledge with which he had no sympathy. The whole great field of modern scientific discovery seemed out of his range. An intimate friend of Faraday once described to me how, when Faraday was endeavouring to explain to Gladstone and several others an important new discovery in science, Gladstone's only commentary was "but, after all, what use is it?" "Why, sir," replied Faraday, "there is every probability that you will soon be able to tax it!"[11]

Unfortunately, Lecky does not identify the "intimate friend of Faraday" who told him the story. Furthermore, Lecky does not specify that the "particular discovery" was in fact the principle of electromagnetic induction.[12]

Faraday's gloss on Franklin—that the answer of the experimenter should be "Endeavour to make it useful"—is well illustrated by Franklin's own scientific career. His invention of the lightning rod certainly made discoveries in electricity useful. On this topic there is some confusion. Because Franklin was a practical man, known for many inventions (including bifocal glasses, an improved stove, and a musical instrument he called the "armonica"), some historians have assumed that his studies of electricity were practical. But when he started out in his scientific career in the 1740s, the only application of electrostatics was in the treatment of certain medical ailments, chiefly paralyses. Although Franklin gave electric shocks to some patients, he did not believe that the electricity itself could have any direct therapeutic value. Rather, with that wonderful insight that was his characteristic, he assumed that what mattered was the patient's will to be cured.[13] Some thirty years later, as a member of a joint committee of the French Royal Academy of Sciences and the Royal Academy of Medicine, Franklin introduced this same shrewd perception to explain the phenomena of mesmerism. There is no mesmeric fluid, the commission concluded, but added in their official report Franklin's observation that the effects seemed to derive from the patient's belief.[14] In 1747, at a moment when some early theorizing had led him astray, Franklin wrote that if "there is no other use discover'd of Electricity, this, however, is something considerable, that it may *help to make a Vain Man humble.*"[15]

Since Franklin did not believe in the only application of electricity then known or envisaged, the motivation for his research was not practical. Rather, like most scientists then and now, he was fascinated by the mysteries of the subject, curious to find out more. Of course, like all good Baconians, and like other scientists of his time, Franklin was convinced that all basic scientific discoveries would eventually lead to some practical embodiment. When he saw the balloon ascensions in 1783, he reported to his friend Joseph Banks, president of the Royal Society, "The Multitude separated, all well satisfied & much delighted with the Success of the Experiment, and amusing one another with Discourses of the various Uses it may possibly be apply'd to, among which were many very extravagant."[16] Franklin's own guess of a possible use was not in relation to technology; rather, he suggesteed that perhaps the new invention "may pave the Way to some Discoveries in Natural Philosophy of which at present we have no conception."

In the 1740s, as his own research progressed, he learned about the "power of points," that is, the effect of points in "drawing off" and "throwing off" the "electric fire." He also studied the role of grounding in electrical experiments. He found out about conductors and insulators and he developed his renowned one-fluid theory of electricity.[17] After studying various forms and conditions of electric discharges, he concluded that lightning is just such an electric discharge as he had been producing in his laboratory. The first test he designed was the experiment of the sentry-box. The significance of this experiment is not usually appreciated. Apart from its importance in validating the theory of lightning as such, and the consequent introduction of the lightning rod, it fundamentally altered the whole nature and status of the subject of electricity.[18]

Prior to these experiments, electrical investigations had consisted of small-scale laboratory experiments in which rods or globes of glass had been rubbed and, to the accompaniment of sparks, other bodies had had the "electric virtue" of the rod transferred to or through them. The German electrical machines had, to be sure, made the sparks bigger, and the Leyden jar had made the effects somewhat greater still so that a shock might actually hurt a human being, or be given to a large number of individuals simultaneously, or even kill a bird. But the subject remained in the area of "toy physics," and was treated as such. Franklin and other "electricians" invented games to make electrical amusements more popular and entertaining; one was called "the electrical mine," a device partially concealed in the floor to shock an unsuspecting person walking over it. Buffon had sponsored the translation of Franklin's book into French because of his appreciation of its genuine scientific worth, but the experiments described in the book were probably performed before the French Court at St. Germain because of their entertainment value.[19]

Within a year of the publication of Franklin's book, however, the situation had been markedly altered. After the sentry-box experiment had been performed successfully in France, and later confirmed in England, every electrician knew thenceforth that the experiments performed with his little laboratory toys might reveal new aspects of one of the most dramatic of nature's catastrophic forces. Electrical studies took on a new importance, and the belief that the operations of the electric fluid might hold the key to important natural processes was partially vindicated. "The discoveries made in the summer of the year 1752 will make it memorable in the history of electricity," William Watson wrote in 1753. "These have opened a new field to philosophers, and have given them room to hope, that what they have learned before in their museums, they may apply, with more propriety than they hitherto could have done in illustrating the nature and effects of thunder; a phaenomenon hitherto almost inaccessible to their inquiries."[20] Electrical phenomena were no longer considered artificial, the result of man's operations, of man's intervention in nature. The lightning experiments showed that electrical phenomena are constantly being produced by natural processes. Thereafter, no natural philosophy could be considered complete if it did not include electricity among the forces of nature.

Franklin's invention of the lightning rod may be seen to have been a practical application of the results of pure, basic, or disinterested research in science. In actual fact, Franklin's invention of the lightning rod was the first major or large-scale example in history of Francis Bacon's prediction that advances in science should lead to practical innovations of benefit to mankind.[21] Almost a century later Franklin's invention of the lightning rod was still being cited as the primary example of the way in which the pursuit of science for purposes of knowledge yields innovations of importance for society.

The history of Franklin's own science-based invention fully exemplifies Faraday's precept that the scientist should endeavor to put his discoveries to use. The baby grew up to become a useful mature adult. Chapin argues persuasively that Franklin seems to have said of the new invention that "It is a child who is just born, one cannot say what it will become," rather than to have asked "What good is a newborn baby?" In both cases, the metaphor was one that came naturally to mind at the time that Franklin saw the early balloon ascensions. As Carl Van Doren pointed out, there actually was at the time a newborn baby in Franklin's household at Passy.[22] This was the two-week-old Ann Jay, daughter of John Jay.

Supplement: The Franklin Stove

Samuel Y. Edgerton, Jr.

Of all of Franklin's inventions and discoveries only the stove has seemed to merit the accolade of an eponym. In the eighteenth century, the lightning rod often bore his name, as in the commemorative tablet in the chapel of Portiuncula, near Assisi, which refers to "electricae Franklinii virgae" or "electrical Franklin rods" (see Chapter 8). The Franklin stove, however, is perhaps the best known of Franklin's inventions.

Those who are familiar with Franklin's career will be aware that Franklin not only wrote about the new stove (or stoves) which he designed, but also wrote extensively about the problem of smoky chimneys, to which he applied the same inverted-siphon principle. It is not generally appreciated, however, that Franklin's innovation went beyond merely placing a free-standing stove so that its heat would not go directly up the chimney. Rather, Franklin drew on his studies of the scientific and technological literature to devise a stove based upon new principles of heat.

The essential features that distinguished the stove (or stoves) which Franklin invented, as Samuel Edgerton has discovered, are not found in any of the free-standing iron fireplaces which today go by the name of "Franklin stove." Later modifications were introduced, presumably in order to make the stove function better than the original design. It is no wonder that Franklin was not wholly successful in designing a fireplace, since the phenomenon of convection was not at that time fully understood. Only after the later work of Rumford could truly efficient stoves or fireplaces be designed.

What is of the greatest interest about Franklin's stoves, in the present context, is that his proposed innovations did not arise from mechanical gadgetry alone but depended on his application of new scientific principles, as was the case for the lightning rod. Both inventions display Franklin's concern for producing simple and economical devices that would improve or make secure the buildings used by ordinary men and women. They exhibit Franklin's concern for the improvement of humanity's estate and his endeavors to put science to use for that end.

I.B.C.

Everyone likes to talk about the "Franklin stove," but few know anything about it. How many owners of what are advertised in antiques journals today as "Franklin stoves" have ever compared their prized possession with Franklin's publications on the subject—particularly his 1744 tract, *An Account of the New Invented Pennsylvanian*

Fire-Places, or his 1785 paper, "Description of a New [Vase] Stove for burning Pitcoal, and consuming all its Smoke"?[1] After taking a careful look at the text, illustrations, and context of these writings, I have concluded that what has since been called by so distinguished a name has had little to do with Franklin's intentions. He devised not just one but four kinds of stoves, all based on an idea, derived from Newton's and Boerhaave's recent theories of matter and heat, that smoke could be made to "descend."[2] Yet in the end, after some fifty years of experimentation, Franklin's altruistic hope of applying this concept to an efficient and economical iron "warming machine," as he called it, was premature and naive. No eighteenth-century heating device that exactly replicates any of Franklin's four original models, nor any reproduction thereof, is extant and workable today.

Nonetheless, because of pioneers like Benjamin Franklin, heat and fire are no longer perceived as supernatural forces beyond human comprehension. Today, we learn in high school that combustion is a chemical and physical process in which oxygen recombines with other elements, producing a form of energy akin to light and electricity called heat. In the twentieth century, we have no trouble understanding how this energy can be conducted through certain materials like metal, how it can then be transferred by convection, and can even radiate between separated materials.

These things seem so self-evident that one would think that people would have observed them even in the Stone Age. The fact is that our ancestors suffered miserably in cold weather right up to the time of Franklin's birth. Their quarters were either blazing hot or drafty cold. They lived in smoke-filled rooms where ventilation and heat control were mostly a matter of chance.

The solution, with the intervention of thinkers like Newton and Franklin, had to await a measuring device, the seemingly trivial graduated thermometer. Not until 1606 was one devised, first by Galileo Galilei of Florence, and then perfected by no less than Ferdinand II de'Medici, Grand Duke of that most fecund of Renaissance cities. Although Florentines have still never learned how properly to heat their houses in winter, they did give the world the first means for objectifying the heretofore mystical and subjective experience of heat. Until the thermometer was invented, no one could be sure just exactly what was meant by "hot" and "cold."

In Franklin's time, natural philosophers believed that heat was a corporeal substance fixed in quantity at Creation and distributed like light throughout the universe. All matter contained a finite amount of it; burning released this material back into the atmosphere where, being lighter than earth, water, and air, it would naturally rise to the "circle of fire" at the outer limits of the cosmos. Natural philosophers called heat a "subtle fluid" which flowed like water and found its level upward rather than down.[3] Newton theorized that all matter consists of microscopic "particles" held in mutual attraction, between which rays of light and heat could penetrate. Newton's Dutch follower Hermann Boerhaave hypothesized further that when heat penetrates air, the latter's particles are driven apart or "rarified." Warmed air, becoming lighter in specific weight, thus rises, while "condensed" cold air replaces it below.[4]

These new theories were first given practical house-warming application in France, in a little book which Franklin acknowledged, *La Mécanique de feu,* written about 1713 by Nicolas Gauger.[5] It opens with a charming picture of the author, trusty thermometer in hand, measuring heat in his spacious house. He senses, for instance, that his cellar is cooler in summer and warmer in winter. With his thermometer, however, he proves that just the reverse is true. Gauger next climbs onto a stool and notes that the temperature near his ceiling is higher than at the floor. Heat therefore must naturally rise, he concludes, allowing cold air to fill in the evacuated space below.

What attracted Franklin to Gauger was that the Frenchman had put these scientific hypotheses to work. He claimed to have designed an open fireplace with hollow iron jambs and a chimney back with an air pocket behind. These hollows all opened to one another but were sealed from the chimney flue and were vented instead into the room. As the fire blazed in the fireplace, Gauger reasoned, the air in the hollows should heat up and rise, then pour into the room through the upper openings. Simultaneously, cold air would enter the fireplace through the lower openings to be warmed and returned to the room in continuous circulation.[6] Because this air flow was protected from the actual fire, no smoke was emitted with it. Gauger even attached a regulator by which the amount of heat coming into the room could be controlled.

Furthermore, Gauger designed a little *soufflet* or "blower" opening through the bottom of the hearth, which allowed a stream of cold air in through a duct from the cellar to play directly on the fire. As we shall see, Franklin incorporated this and more of Gauger's ideas into his 1744 Pennsylvanian fireplace.

Gauger advocated an improvement which Franklin chose at first to ignore. This was to take advantage of the Newtonian hypothesis that heat radiates and reflects just like light. Gauger observed that heat travels in the form of rectilinear "rays." When encountering a surface, they reflect at an angle equal to the angle of incidence.[7] The Frenchman claimed to have shaped the back and jambs of his fireplace like a concave

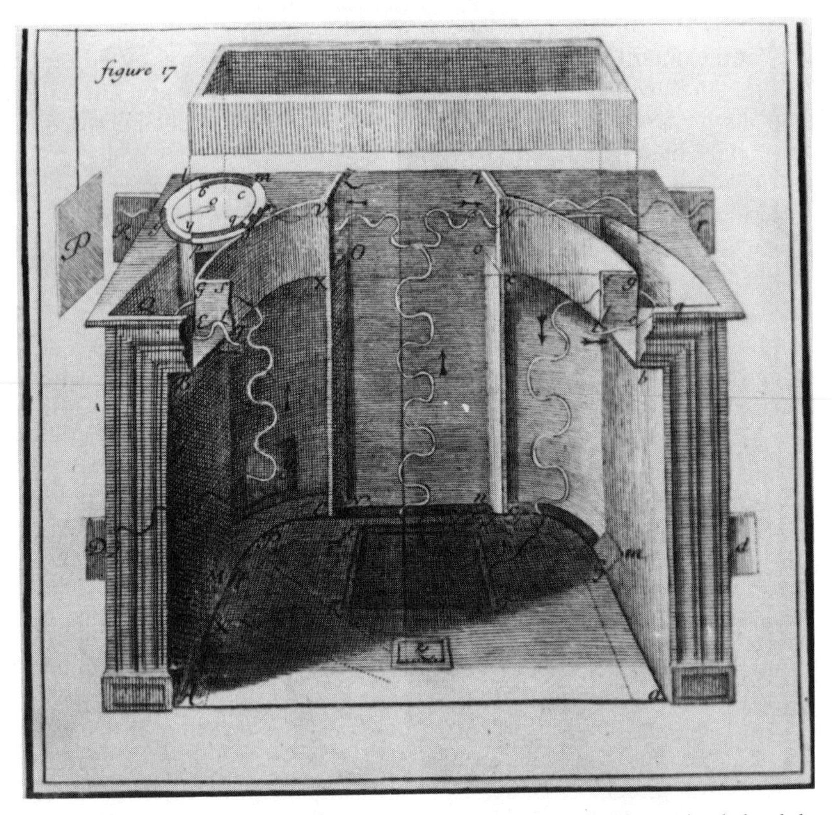

Cutaway view of Gauger's fireplace. Air is shown flowing into the pocket behind the fireplace through the lower openings (*D* and *d*), where it is warmed by the fire and returned into the room through the upper openings (*R* and *r*). Widener Library, Harvard University.

parabola. He even diagrammed geometrically how heat rays, even if they strike his curving fireplace surface obliquely, are reflected straight back into the room. Although Franklin well knew that "rays of heat" do move in "right lines" in all directions from the fire source, he was so eager to apply his "Aerial Syphon" theory that he overlooked what might have been a saving improvement of his 1744 stove.

Franklin was especially motivated to apply Gauger's ideas for reasons of public health. He was much worried, as were all the colonists, about the strenuous winters of North America and the danger of cold drafts that caused so many diseases (often to women "as they sit much in the House"). Cold drafts, he complained in his 1744 pamphlet, were actually caused by current house-warming practice. A roaring fire in the customary large open fireplace left a "Man to be scorch'd before while he's froze behind." Particularly lethal were the streams of frigid air sucked in through uninsulated chinks and cracks in the walls. "If the Wind blows on you thro' a Hole, Make your Will, and take Care of your Soul," Franklin quoted from a Spanish proverb. More seriously, he added a lengthy footnote citing the warnings, some in Latin, of all the respected medical authorities, Oriental as well as Occidental.[8]

In that regard, perhaps the most stimulating book Franklin was reading at the time was by another Frenchman, Jean Théophile Desaguliers, entitled *A Course in Experimental Philosophy.*[9] This work had been purchased by the Library Company of Philadelphia in 1744. Desaguliers was an expatriate French Protestant, an ardent Newtonian and one of the founders of Freemasonry. He had also translated Gauger's *La mécanique du feu* under the English title *Fires Improved . . .*, no doubt Franklin's original source for this work since a copy of it too was in the Library Company, given by Franklin's friend, the ironmaster Robert Grace.[10] Desaguliers added a long postscript to his *Course in Experimental Philosophy,* in which he described Gauger's fireplace and his own modified "warming machines," designed he said for the British Houses of Parliament.[11] What struck Franklin concerning Desaguliers's emendations was that he had thought to substitute cast iron for masonry, even conducting experiments on the "salubriousness" of heated metal. For example, he had placed a small living bird in a glass jar, closed except for an air tube through a heated metallic cube. When the cube was of iron, Desaguliers stated, the bird showed no ill effects whatever, but when brass was substituted the bird died instantly.[12]

Franklin was delighted with this experiment, as he recorded in his own 1744 tract. The Philadelphian had a vested interest in iron, which was one of the largest industries in Pennsylvania. Franklin's good friend Robert Grace was the proprietor of the prosperous Warwick Furnace in Chester County. On the flyleaf of his 1744 pamphlet, Franklin printed an "Advertisement" awarding Grace exclusive rights to manufacture the Pennsylvanian fireplace. Franklin's tract served as an advertisement for the industry in general. As he remarked, "Iron is always sweet, and every way taken is wholesome and friendly to the human body—except in Weapons."

It is necessary to explain that Franklin was not alone in the house-warming business in eighteenth-century America.[13] Many kinds of iron stoves were manufactured and sold in the Colonies; Franklin specifically referred to several of these in his pamphlet. Perhaps the most popular for warming large assembly halls and the like was called by Franklin the "Holland Stove," but more commonly the "six-plate stove." It consisted of an iron box of six separate metal sides bolted together and raised off the floor on short legs. A hinged door on one end allowed fuel to be inserted, and a hole in the top plate permitted the smoke to rise through a pipe.

Franklin had also observed a unique variation among the German settlers of rural Pennsylvania. Like the English, the Germans built their houses with central chimneys and large open fireplaces, but with an added open-ended box made of five cast-iron plates. The open end was fitted into the back of the chimney so that the box could extend into an abutting bedchamber. Before retiring, the German householders would shovel the still-hot coals from the fire on the kitchen side of the chimney into the iron stove on the other. The heat from the coals would then be convected into the bedchamber.

Five- and six-platers were known as "closed stoves," meaning that the fire inside was unexposed to view. Closed stoves with their multiple convecting surfaces had the obvious advantage of producing more heat. They originated in continental northern Europe, in Germany, Holland, Scandinavia, and Russia. The English preferred the coziness of the large open hearth in spite of its voracious appetite for firewood. They remained habitually suspicious of a "fyre secret felt but not seene." Benjamin Franklin, clever merchandiser as well as ingenious natural philosopher, decided to combine the fuel-saving efficiency of the closed iron stove with the romantic open fireplace loved by the English.

But scientific facts, Franklin asserted, were the true guiding principles of his Pennsylvanian fireplace: "Heat may be separated from the Smoke as well as from the Light, by means of a Plate of Iron, which will suffer Heat to pass through it without the others . . . [Smoke] is but just separated from the Fuel; and then moves only as it is carried by the Stream of rarified Air."[14] Moreover, he was convinced it was feasible to force some of the rarified air bearing the smoke downward below the fire itself, through covered ducts in the floor leading to the flue. Franklin assumed, based on his reading of Boerhaave, Desaguliers, and Gauger, that because hot air rises just as ordinary liquids descend, he could create what he called an "Aerial Syphon" or "Syphon revers'd"—that is, a draft strong enough to carry the smoke downward and horizontally through ducts under the floor before releasing it up the chimney. For the rest of his life, even while living abroad in England and France, Franklin was to be preoccupied with this idea. Here is how he explained it some years later, in regard to his "vase" stove design, in a letter to the Marquis de Turgot:

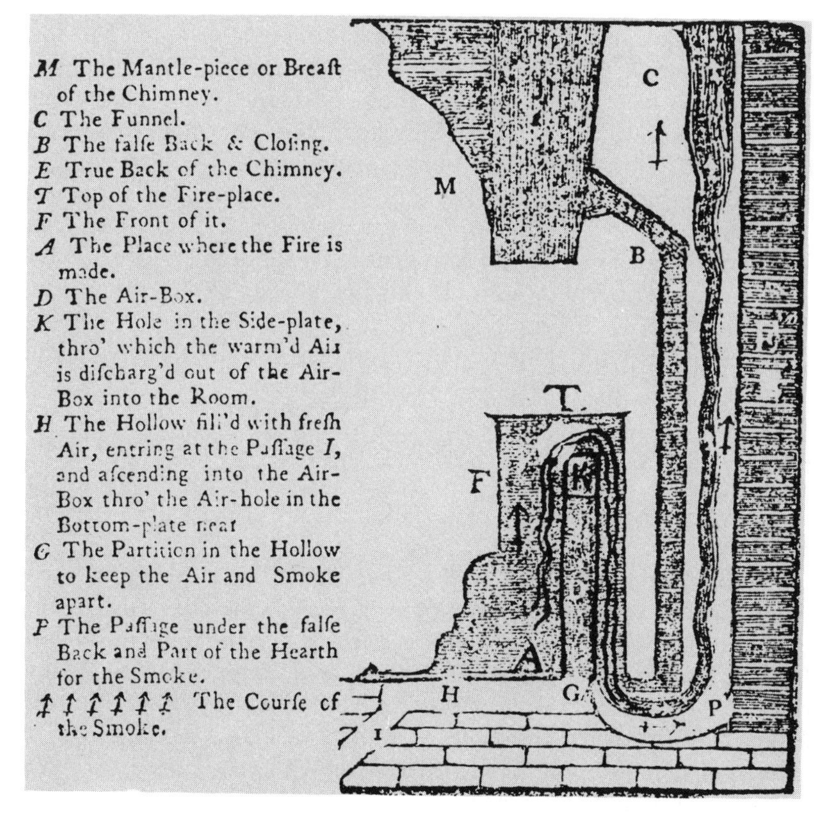

Operation of the Pennsylvanian fireplace, from Franklin's 1744 tract. The enclosed air box (*D*) draws in fresh air through a hole in the bottom plate (*G*), where it is warmed and vented back into the room through holes in the side plates (*K*). The smoke flows over the air box, down through a passage in the floor (*P*), and up the flue. Chapin Library of Rare Books, Williams College.

> Its Principle is that of a Syphon revers'd, operating on Air in a manner somewhat similar to the Operation of the common Syphon on Water. The Funnel of the Chimney is the longer Leg. The Vase [upper body of the stove] is the shorter. And as in the common Syphon, the Weight of Water in the longer Leg is greater than in the shorter Leg, and in thus descending permits the Water in the shorter Leg to rise, by the Pressure of the Atmosphere: So in the Aerial Syphon, the Levity of the Air in the longer Leg being greater than that in the Shorter, it rises & permits the Pressure of the Atmosphere to force that in the Shorter to descend. This causes the Smoke to descend also . . ."[15]

Let us look at Benjamin Franklin's "New Invented Pennsylvanian Fire-Place" as he described its assemblage and operation in the 1744 pamphlet.[16] While Gauger and Desaguliers had thought to apply Newton's and Boerhaave's theories of heat to house-warming, they offered no advice as to how such fireplaces could be economically constructed. Franklin therefore had to contrive his stove from scratch, in effect learning the sand-casting technology of iron manufacture. The stove was to be about thirty inches high, consisting of a top, two side plates, a bottom plate, and a decorated panel covering the upper part of the open front. The back plate was to stand a few inches in front of the bricked-up flue, leaving the only openings into the chimney by way of the bottom plate. Franklin observed conventional six-plate stove construction by bolting the sides together with long wrought-iron "screw rods" through holes in pre-cast "ears" at the edges of the top and bottom plates. His most unconventional feature was the "air box," consisting of two plates with extended baffles set perpendicularly between the side plates. Together they formed a separate enclosure inside the stove with openings only through holes in the bottom plate and in the upper parts of the two side plates.

The fire in the open part of the stove heated the baffled iron plates, which in turn warmed the smoke-free air inside the box. This air then circulated around the baffles and was released into the room through the holes in the side plates. Simultaneously, cold air pressed into the box through a hole in the bottom plate. Franklin, following Gauger, attached a little brass door to this hole as a "blower" to direct some of the incoming fresh air upon the fire itself. Meanwhile, the smoke, carried by whatever heated air remained inside the stove, must rise (by "Law of Nature" as Franklin said). Its only exit to the chimney, however, led over and *down* behind the air box, through another set of holes in the bottom plate, under the back of the stove, and out a passage dug in the floor leading to the flue.[17]

The householder, Franklin advised, must find a good mason to help set up this immensely heavy apparatus within the already existing fireplace. It was essential that the entire Pennsylvanian fireplace stand free, so that all of its iron plates would convect heat only into the room and none of it would be wasted up the chimney.

Like Henry Ford, Benjamin Franklin envisioned his invention in mass production, cheap and conveniently available to the common man. To symbolize that end, he designed a logo for the front plate, a beaming sun and the words ALTER IDEM, the meaning of which he explained in his pamphlet along with a rhyming commercial:

> On the Device of the NEW FIRE-PLACE, A SUN; with this Motto, ALTER IDEM. i.e., A Second Self; or, Another, the same. By a Friend.

ANOTHER Sun!—'tis true; but not THE SAME.
Alike, I own, in Warmth and genial Flame:
But, more obliging than his elder Brother,
This will not scorch in Summer, like *the other;*
Nor, when sharp *Boreas* chills our shiv'ring Limbs,
Will *this Sun* leave us for more Southern Climes;
Or, in long Winter Nights, forsake us here,
To chear new Friends in t'other Hemisphere:
But, faithful still to us, this *new Sun's* Fire,
Warms when we please, and just as we desire.

In the 1744 pamphlet, Franklin claimed his invention had already given satisfaction to a "great Number of Families in Pennsylvania" for "now three Winters." Apparently, in early 1741 he was advertising in the *Pennsylvania Gazette* "Very good Iron Stoves," and later in the same year "New Invented Fire-Places" for five pounds each.[18] We know also that he appointed two of his brothers and several friends as agents for marketing the stove in a number of cities in the Colonies. According to Franklin's own estimates, business was brisk.[19] Franklin's pamphlet was widely distributed by the author, and even translated into Dutch, German, Italian, and French, eliciting admiring testimonials from politicians and fellow natural philosophers in Europe as well as America. For the rest of his life, Franklin's opinion was solicited on all sorts of practical and theoretical house-warming problems. To this day scholars have never doubted that the Pennsylvanian fireplace performed exactly as the Philadelphia doctor claimed, and that it was a success.

But was it? We should expect glowing testimonials by actual users; curiously, there are none (except by Franklin himself). In fact, close examination of the commercial records shows that business was actually bad. In the year the Pennsylvanian fireplace pamphlet appeared, Peter Franklin, the inventor's brother and special agent for marketing the device in Newport, Rhode Island, ordered eleven stoves wholesale from the Pennsylvania manufacturer. For the next twenty years, Peter Franklin's ledger books, now preserved in the American Philosophical Society Library in Philadelphia, show only two retail sales.

Moreover, manufacturing of the stove had all but ceased before two decades had passed. On 9 August 1765, Benjamin Franklin wrote from London to the Quaker merchant Hugh Roberts of Philadelphia asking that two of the stoves be procured and shipped to him by the next boat. He added, "As many People laid them aside, I hope this will not be difficult." It was. On November 27 of the same year, Roberts replied:

Thy request of procuring 2 Pennsylvania Fire Places cast when Robert Grace's Moulds were good, is a little uncertain which of the sorts thou

The sole surviving Pennsylvanian fireplace. Note the solid bottom piece and the hole cut in the back plate. Mercer Museum of the Bucks County Historical Society, Doylestown, Pa.

> intended, whether the first impression such as P Syng and I had or that with the Sun in front and Air Box, both of which are much out of use, and tho' many have been laid aside I find on enquiry that some parts of the plates have been apply'd for Backs or hearths of Chimneys or other Jobs, that I have not found a second hand one Compleat.[20]

Perhaps the most compelling testimony to how well the Pennsylvanian fireplace worked is provided by the single surviving example. In size and outward appearance at least, this rusting relic, including the ALTER IDEM logo, is quite similar to Franklin's original engraving. However, the baffled internal plates which once formed three sides of the air box are missing. In fact, this feature, so basic to Franklin's design, has been altogether eliminated; the bottom plate has been cast as a solid piece, without the holes for drawing air up from below the floor into the air box and without the holes for the smoke to go down. Instead, the stove has been modified so that the smoke will escape upward by way of the back plate, through a jagged rectangular hole cut through the cast iron. For the smoke to pass out through this aperture, the back of the stove would have to be sealed inside the masonry fireplace under the flue, instead of standing free in the room as Franklin advocated.

In sum, this sole surviving Pennsylvanian fireplace has been remodeled beyond any reconciliation with Franklin's intentions. Except for exhibiting its fire, the operating principle here is little different from that of a conventional six-plate stove, with the disadvantage that much of its convecting surface would have been bricked into the chimney wall. That this mutilated artifact exists, while the prototype has totally disappeared, is mute evidence that Franklin's 1744 pamphlet promised more than the actual product delivered.

The design of the Pennsylvanian fireplace had at least one fatal deficiency: the manner in which the smoke was to be forced down under the back plate and up into the chimney. As Franklin well knew, his "Syphon revers'd" effect could take place only if the walls of the smoke-egress remained *uniformly warmer* than the outside surrounding air. The internal air within this passage would then increasingly "rarify" and seek an upward exit, conveying away the heavier smoke. The glitch was the section of the passage imbedded in the masonry hearth between the iron stove and the chimney. In order for the smoke to be forced down and along this, its brick walls would have to be exceedingly hot. Unless the householder took special care to keep the fire constantly fueled, the brick passage would tend to cool. The smoke upon reaching this point would then find the draft diminished and spill back into the room.

Franklin was clearly hurt by the poor reception of his Pennsylvanian fireplace. Frequently in his letters during the rest of his life he complained that its operation had not been properly attended to. As late as 1785 he was still berating the many "disfigured . . . imitations," insisting that "one of its main intentions . . . of admitting a sufficient quantity of fresh air warmed in entering through the air-box" had been "nearly defeated, by a pretended improvement."[21] Interestingly, even the name "Pennsylvanian fireplace" quickly lost its currency. By the early 1770s, expurgated versions of Franklin's invention, advertised as "open 6 plate half stoves" or "open stoves with flue," were being manufactured and sold in Philadelphia.[22] Characteristic of all these ersatz examples were a smoke outlet near the top and total elimination of the air box.

By the 1780s, the most common of these modified versions came to be known as the "Rittenhouse stove," because of specific alterations introduced presumably by David Rittenhouse, the American astronomer and natural philosopher who also resided in Philadelphia.[23] Many eighteenth-century examples of the Rittenhouse-type stove are extant today.

Franklin's 1773 engraving of the vase stove. The fire was contained in the cast-iron box below the urn, sitting on a larger box that contained air baffles reminiscent of the 1744 stove design. The heated air was drawn down into the baffle box, where its heat was convected to the room, and then into the flue. American Philosophical Society, Philadelphia.

This is the stove most commonly and erroneously referred to as the "Franklin stove."

Franklin persisted in his experiments with fireplace improvements and stove designs during his years abroad, including an ornamental coal-burning "vase" stove based on principles similar to those of the Pennsylvanian fireplace. After he returned to the United States in 1785, he continued to write and lecture about chimney design, fireplace improvements, and the vase stove.[24] But the fate of these inventions was even more discouraging than that of the Pennsylvanian fireplace. There is no evidence that the vase stove, or any of the fireplace devices he had been experimenting with since the 1750s, ever went into production. Nonetheless their designs were frequently plagiarized and modified by others without credit to their originator.[25]

In truth, during the last five years of his life, Benjamin Franklin, in spite of being an internationally recognized authority on matters of house-warming, was a prophet without honor in his own country. One wag even went so far as to suggest the vase stove was best suited as a monument for the great man's grave, inscribed with this verse as epitaph:

Let candor then write on this urn,
Here lies the renowned inventor
Whose fame to the skies ought to burn
But inverted descends to the center.[26]

For the next fifty years, Franklin's followers—excellent technicians, to be sure, such as Rittenhouse, Charles Willson Peale, and Count Rumford—did make significant practical improvements on his initial ideas. Too often, in their eagerness to capitalize on these modifications, they conveniently forgot to cite his pioneering efforts. It is ironic that the fickle muse of history decided to attach Franklin's name to every kind of house-heating device except the ones he actually invented. In the long run, however, it must be said that his true contribution to the subject had more to do with promotion than production. As Sydney George Fisher, one of his biographers, remarked, the proof of Benjamin Franklin's universal genius was his uncommon ability to be absolutely fascinating even in writing about stoves.

Notes

1. *Introduction*

1. *Benjamin Franklin's Experiments: A New Edition of Franklin's Experiments and Observations on Electricity,* edited, with a critical and historical introduction, by I. Bernard Cohen (Cambridge, Mass.: Harvard University Press, 1941). I have prepared a facsimile reprint of the first edition of Franklin's *Experiments and Observations* with a historical introduction and critical appendices, to be published by the Dibner Library in Washington, D.C., in association with the Smithsonian Press.
2. The historian in question was Curtis P. Nettels; see Chapter 3.
3. Ferd. Rosenberger, *Die moderne Entwicklung der elektrischen Principien* (Leipzig: Johann Ambrosius Barth, 1898), pp. 16–24; Edmund Hoppe, *Geschichte der Elektrizität* (Leipzig: Johann Ambrosius Barth, 1884), pp. 27–42, "Das Zeitalter Franklins und Coulombs 1747–1789. Franklin und seine Zeitgenossen."
4. Mario Gliozzi, *L'elettrologia fino al Volta* (Naples: Luigi Loffredo, 1937), vol. 1, pp. 148–208.
5. Brother Potamian and James J. Walsh, *Makers of Electricity* (New York: Fordham University Press, 1909), pp. 68–132. Park Benjamin, *A History of Electricity (The Intellectual Rise in Electricity) from Antiquity to the Days of Benjamin Franklin* (New York: John Wiley & Sons, 1898; rept. New York: Arno Press, 1975), pp. 537–593. Reference should also be made to Dayton Clarence Miller, *Sparks, Lightning, Cosmic Rays: An Anecdotal History of Electricity* (New York: Macmillan, 1939), based on a series of "Christmas Week Lectures for Young People" given in 1937 at the Franklin Institute.
6. Paul Leicester Ford, *The Many-Sided Franklin* (New York: Century, 1899), pp. 351–387; Edwin J. Houston, "Franklin as a Man of Science and an Inventor," *Journal of the Franklin Institute,* 161 (1906), 241–316, 321–383. Ford devotes only a very small part of his chapter to Franklin as a founder of electrical theory.

7. Dorothy M. Turner, *Makers of Science: Electricity and Magnetism* (London: Oxford University Press, 1927), pp. 24–27. To these should be added E. T. Whittaker, *A History of the Theories of Aether and Electricity from the Age of Descartes to the Close of the Nineteenth Century* (Dublin: Dublin University Press, 1910). This work was revised, rewritten, and expanded into a two-volume work entitled *A History of the Theories of Aether and Electricity;* vol. 1, subtitled "The Classical Theories" (Edinburgh and London: Thomas Nelson; New York: Philosophical Library, 1951), contains a section (pp. 46–51) entitled "The One-Fluid Theory: Ideas of Watson and Franklin." See also the book by Mottelay cited in Note 19, and especially Priestley's history, cited in Note 21.
8. This subject is developed in my *Franklin and Newton: An Inquiry into Speculative Newtonian Experimental Science and Franklin's Work in Electricity as an Example Thereof* (Philadelphia: American Philosophical Society, 1956; Cambridge, Mass.: Harvard University Press, 1966). See also Chapter 2.
9. These issues are discussed in my article "Science and the Growth of the American Republic," pp. 67–106 in Walter Nicgorski and Ronald Weber, eds., *An Almost Chosen People* (Notre Dame, Ind.: University of Notre Dame Press, 1976).
10. Historians of science then obtained their doctorates in departments of history, or philosophy, or even literature. A notable group of historians of science who did their graduate work at Columbia in the 1930s were there classified as historians in the Graduate Faculty of Political Science; this group included Carl B. Boyer, C. Doris Hellman (Pepper), Pearl Kibre, Edward Rosen. So far as I know, the first doctorate in the history of science (as a separate and recognized subject) in the United States was awarded to Aydin M. Sayili (Harvard, 1942); the second was awarded to the present writer (Harvard, 1947), the edition of *Benjamin Franklin's Experiments* being accepted as a doctoral dissertation (see note 1).
11. See, for example, vol. 1 (1985) of the newly reconstituted journal *Osiris,* edited by Sally Kohlstedt and Margaret Rossiter, dealing with "Historical Writing on American Science." See also *History of Science in America: News and Views,* Newsletter of the Forum for the History of Science in America, ed. Clark A. Elliott (University Archives, Harvard University).
12. Carl Van Doren, *Benjamin Franklin* (New York: Viking Press, 1938; New York: Garden City Publishing Co., 1941), pp. 156–182.
13. Ronald W. Clark, *Benjamin Franklin: A Biography* (New York: Random House, 1983).
14. Esmond Wright, *Franklin of Philadelphia* (Cambridge, Mass.: Harvard University Press, 1986).
15. See Cohen, *Franklin and Newton,* esp. chap. 8.
16. The quotation is from a report on electricity written from Paris by Turberville Needham, "Extract of a Letter . . . concerning Some New Electrical Experiments lately made at Paris," *Philosophical Transactions,* 44 (1746), 247–263. Needham adds that Buffon held the same opinion.

17. In a note added to the first epistolary report on his research (Philadelphia, 25 May 1747), Franklin said: "This power of points to *throw off* the electrical fire, was first communicated to me by my ingenious friend Mr *Thomas Hopkinson* . . ." See Cohen, *Benjamin Franklin's Experiments,* p. 172.
18. A bibliographical description of the several editions appears as chap. 4, sec. 1 of Cohen, *Benjamin Franklin's Experiments* (see Note 1).
19. Among those who repeated the kite experiment were John Lining in America and Jacques de Romas in France. See the chronological table on p. 320 of Paul Fleury Mottelay, *bibliographical History of Electricity Chronologically Arranged* (London: Charles Griffin, 1922; Philadelphia: Lippincott, 1922); see also Chapter 6.
20. There are many problems concerning the background of the discovery or invention of the Leyden jar, and its dissemination; see John L. Heilbron, "A propos de l'invention de la bouteille de Leyde," *Revue d'Histoire des Sciences et de leurs Applications,* 19 (1966), 133–142; J. L. Heilbron, "G. M. Bose: The Prime Mover in the Invention of the Leyden Jar?" *Isis,* 51 (1966), 264–267; C. Dorsman and C. A. Crommelin, "The Invention of the Leyden Jar," *Janus,* 46 (1957), 275–280; Jean Torlais, "Qui a inventé la bouteille de Leyde?" *Revue d'Histoire des Sciences et de leurs Applications,* 16 (1963), 211–219. For Franklin's analysis of the action of the Leyden jar, see Chapter 2.
21. Joseph Priestley, *The History and Present State of Electricity,* 3rd ed. (London, 1775; facsimile rept., New York and London: Johnson Reprint Corporation, 1966, with a valuable introduction by Robert E. Schofield), vol. 1, p. 107. Priestley's *History* is still a most important source of information concerning Franklin's work in electricity and aspects of eighteenth-century research in electricity, although it is weak on certain Continental developments and has an obvious pro-Franklin bias (see Note 33).
22. As William Watson put it: "The discoveries made in the year 1752 [with respect to lightning and the electrification of clouds] have opened a new field to philosophers, and have given them room to hope, that what they have learned before in their museums, they may apply, with more propriety than they hitherto could have done, in illustrating the nature and effects of thunder; a phaenomenon hitherto almost inaccessible to their inquiries." See W. Watson, "An Account of a Treatise . . ." *Philosophical Transactions,* 48 (1753), 201–202.
23. Franklin soon discovered that he could detect the electrification of clouds even when there was no stroke of lightning; see Chapters 6 and 7.
24. I owe this reference to Professor L. Pearce Williams of Cornell University.
25. Franklin to Cadwallader Colden, Philadelphia, 20 Sept. 1748.
26. Van Doren, *Benjamin Franklin,* p. 187.
27. Banks to Franklin, London, 25 Aug. 1783, Cohen, *Benjamin Franklin's Experiments,* p. 8.
28. Franklin to Banks, Passy, 9 Sept. 1782.
29. Franklin to Ingenhousz, Passy, 20 Apr. 1785.

30. The full quotation reads: "Eripuit coelo fulmen sceptrumque tyrannis" (He snatched the lightning from the sky and the sceptre from tyrants). See *Journal de Paris,* 15 Nov. 1778, and Cohen, *Benjamin Franklin's Experiments,* p. xxvii.
31. Needham, "Extract of a Letter . . ."
32. See Cohen, *Benjamin Franklin's Experiments,* chap. 4.
33. On this aspect of the introduction of Franklin's book into France, see John L. Heilbron, "Franklin, Haller, and Franklinist History," *Isis,* 68 (1977), 539–549. Heilbron points out that Nollet's theory was not so completely demolished by Franklin as Franklin alleged (chiefly in his autobiography), and as Priestley (following Franklin) unequivocally declared, and as I too indicated (in my *Benjamin Franklin's Experiments* and again in *Franklin and Newton,* where my chief aim was to study Franklin's ideas rather than those of his opponents). In particular, Heilbron contrasts the reception of Franklin's one-fluid theory in England and in France. This aspect of history is explored at length in J. L. Heilbron, *Electricity in the 17th and 18th Centuries: A Study of Early Modern Physics* (Berkeley: University of California Press, 1979), chap. 15. For more on Nollet, see Isaac Benguigui, "La théorie de l'électricité de Nollet et son application en médecine à travers sa correspondance inédite avec Jallabert," *Gesnerus,* 38 (1981), 225–235; Benguigui, ed., *Théories électriques du XVIII siècle: Correspondance entre l'Abbé Nollet (1700–1770) et le physicien genévois Jean Jallabert (1712–1768)* (Geneva: Georg, 1984); R. W. Home, "Nollet and Boerhaave: A Note on 18th Century Ideas about Electricity and Fire," *Annals of Science,* 36 (1987), 171–176; Elizio Yamazaki, "L'Abbé Nollet et Benjamin Franklin. Une phase finale de la physique Cartésienne: la théorie de la conservation de l'électricité et de l'expérience de Leyde," *Japanese Studies in the History of Science,* 15 (1976), 37–64.
34. Translated from Jacques-Barbeu-Dubourg's comments in his French edition of Franklin's book on electricity, vol. 1, pp. 335–338.
35. Giambattista Beccaria, *A Treatise upon Artificial Electricity . . .* (London, 1776), pp. 1, 18; for more on Beccaria, see Heilbron, *Electricity in the 17th and 18th Centuries,* pp. 362–372, and the article on Beccaria in the *Dictionary of Scientific Biography.*
36. J. J. Thomson, *Recollections and Reflections* (London: Bell, 1936), pp. 252–253.
37. See Cohen, *Franklin and Newton,* pp. 452–467, 516–533. For Franklin's analysis, see Chapter 2. The charging of the Leyden jar as a result of bringing the coating rather than the hook (or knob) to a rubbed globe or tube—as Franklin said—was "a discovery of the ingenious Mr. [Ebenezer] Kinnersley, and by him communicated to me." See J. A. Leo Lemay, *Ebenezer Kinnersley: Franklin's Friend* (Philadelphia: University of Pennsylvania Press, 1964).
38. Franklin acknowledged that the phenomenon of the repulsion of two negatively charged bodies had been discovered by Ebenezer Kinnersley; see Lemay, *Ebenezer Kinnersley.*

39. Aepinus, *Tentamen theoriae electricitatis et magnetismi* . . . (St. Petersburg: Academia Scientiarum, 1759). For an English translation, as well as a critical introduction and extensive commentary, see R. W. Home and P. J. Connor, *Aepinus's Essay on the Theory of Electricity and Magnetism* (Princeton: Princeton University Press, 1979).

In my presentation of "Aepinus's Revision of the Franklinian Theory" (pp. 537–543 of *Franklin and Newton*), I misleadingly stressed the failure of the Franklinian theory to explain or to account for the mutual repulsion of negatively charged bodies as the primary reason for Aepinus's introduction of the postulate that "common matter" deprived of its normal component of electrical fluid will repel similar common matter. As Heilbron has pointed out, Aepinus "introduced the additional hypothesis that the particles of common matter are mutually repulsive . . . with reference to minus-minus repulsion, not as an ad hoc minor adjustment of the theory, but as an essential ingredient without which none of Franklin's analysis could stand. For, as he rightly observed, without repulsion between matter particles, uncharged neutral bodies would attract one another electrically." That is, "Although the forces on the electrical fluid in each of the interacting bodies sum to zero, the matter in each feels an uncompensated attraction from the fluid of the other." This deficiency in the Franklin theory was noted by Gliozzi, *L'elettrologia,* vol. 1, pp. 188–189, and by Rosenberger, *Entwicklung,* p. 28; Home discusses the issue in Home and Connor, *Aepinus's Essay,* pp. 118–121.

40. See William Whewell, "Report on the Recent Progress and Present Condition of the Mathematical Theories of Electricity, Magnetism, and Heat," *Report of the Fifth Meeting of the British Association for the Advancement of Science* (Dublin, 1835), pp. 1–34. See also Cohen, *Franklin and Newton,* pp. 543–554; Heilbron, *Electricity,* chap. 18.

41. See Heilbron, *Electricity,* chap. 15, sec. 3.

42. See Roderick W. Home, "Franklin's Electrical Atmospheres," *British Journal for the History of Science,* 6 (1972), 131–151; *The Effluvial Theory of Electricity* (New York: Arno Press, 1981).

43. Roderick W. Home, "Introductory Monograph," in Home and Connor, *Aepinus's Essay.*

44. The two-fluid theory was put forward by Robert Symmer, a neglected figure in the history of electricity and of eighteenth-century science; see John L. Heilbron, "Robert Symmer and the Two Electricities," *Isis,* 67 (1976), 7–20. See also Cohen, *Franklin and Newton,* pp. 543–546.

45. See, for example, Robert Andrews Millikan, "Benjamin Franklin as a Scientist," *Journal of the Franklin Institute,* 2 (1941), 407–423, esp. 417; this essay was reprinted in a volume of essays on Franklin, with a foreword by Henry Butler Allen, entitled *Meet Dr. Franklin* (Philadelphia: Franklin Institute, 1943), in which the reference to Franklin and the electron occurs on p. 21. Millikan writes: "Franklin states with great succinctness what later became known as the Franklin one-fluid theory, and after 1900 was known as the electron theory." See also Millikan's *The Electron* (Chicago: Univer-

sity of Chicago Press, 1917), pp. 19, 24, and his article "Franklin's Discovery of the Electron," *American Journal of Physics,* 16 (1948), 319.

46. C. B. W. [C. B. Wilde], "Whig History," pp. 445a–446a in W. F. Bynum, E. J. Browne, and Roy Porter, eds., *Dictionary of the History of Science* (Princeton: Princeton University Press, 1981).

2. *Franklin's Scientific Style*

1. The role of Newton's *Opticks* as a manual of experimental art in the eighteenth century is explored at length in my *Franklin and Newton* (Philadelphia: American Philosophical Society, 1956; Cambridge, Mass.: Harvard University Press, 1966). Newton's *Opticks* was published in London in 1704; a facsimile reprint was issued in Brussels in 1966 by Editions Culture et Civilisation. A reprint of the fourth edition (London, 1730), with a foreword by Albert Einstein, an introduction by Sir Edmund Whittaker, and an analytical table of contents prepared by Duane H. D. Roller, was issued by Dover Publications (New York, 1952; rev. ed., 1979).
2. While I was studying the seminal role of the *Opticks* for electrical scientists and other experimenters, Marjorie Hope Nicolson was exploring the influence of science on the eighteenth-century literary imagination and found that poets (and others) had been influenced in a unique way by the *Opticks* and not by the *Principia.* See her *Newton Demands the Muse: Newton's "Opticks" and the Eighteenth-Century Poets* (Princeton: Princeton University Press, 1946). Important additions to our understanding were made by Robert E. Schofield, *Mechanism and Materialism: British Natural Philosophy in an Age of Reason* (Princeton: Princeton University Press, 1970) and by Richard Olson, *Scottish Philosophy and British Physics, 1750–1880: A Study in the Foundations of the Victorian Scientific Style* (Princeton: Princeton University Press, 1975). For a brief survey of more recent scholarship see Simon Schaffer, "Physical Science," pp. 285–314 in Pietro Corsi and Paul Weindling, eds., *Information Sources in the History of Science and Medicine* (London: Butterworth Scientific, 1983).
3. Preface to the first edition (1687) of the *Principia: "Quo sensu* mechanica rationalis *erit scientia motuum . . . accurate proposita ac demonstrata."* (In this sense rational mechanics will be the science of motions . . . accurately proposed and demonstrated.)
4. In experiment 15, accompanying proposition 6, theorem 5, of book 1, part 1, Newton refers to "the Proportions of the Sines" which "come out equal, so far as by viewing the Spectrums, and using of some mathematical Reasoning I could estimate." This uncharacteristic use of mathematics requires a proposition stated as follows: "If any Motion or moving thing whatsoever be incident with any Velocity on any broad and thin space terminated on both sides by two parallel Planes, and in its Passage through that space be urged perpendicularly towards the farther Plane by any force which at given distances from the Plane is of given Quantities; the perpendicular velocity of that Motion or Thing, at its emerging out of that space, shall be always

equal to the square Root of the sum of the square of the perpendicular velocity of that Motion or Thing at its Incidence on that space; and of the square of the perpendicular velocity which that Motion or Thing would have at its Emergence, if at its Incidence its perpendicular velocity was infinitely little." Newton then observes that "the same Proposition holds true of any Motion or Thing perpendicularly retarded in its passage through that space, if instead of the sum of the two Squares you take their difference." This leads him to the remarkable statement that the "Demonstration" is one which "Mathematicians will easily find out, and therefore I shall not trouble the Reader with it." Most mathematicians, I suspect, would most easily find out the demonstration by turning to *Principia,* book 1, section 14, proposition 94.

5. *Opticks,* book 2, part 2 (p. 240 of Dover edition).
6. The absence of formal mathematics in the printed text of the *Opticks* would have been aggrandized for readers of the first edition by the presence of mathematics in the appendix. This consisted of two mathematical tracts by Newton in Latin: *Enumeratio Linearum Tertii Ordinis* and *Tractatus de Quadratura Curvarum* (both presented without the author's name). The latter contained the first full-scale formal presentation in print by Newton of his methods of the differential and integral calculus. Here Newton displayed in print for the first time his algorithm based upon the use of dotted or "pricked" letters (x with one dot above it for dx/dt; x with two dots above it for d^2x/dt^2; x with three dots above it for d^3x/dt^3; and so on) and gave a table of integrals, together with a presentation of his methods of integration.
7. Zev Bechler, "Newton's Search for a Mechanistic Model of Colour Dispersion," *Archive for History of Exact Sciences,* 11 (1973):1–37; "Newton's Law of Forces which are Inversely as the Mass: A Suggested Interpretation of His Later Efforts to Normalize a Mechanistic Model of Optical Dispersion," *Centaurus,* 8 (1974): 184–222. Alan E. Shapiro, "Experiment and Mathematics in Newton's Theory of Color," *Physics Today,* 37 (1984): 34–42.
8. It would not have been hard to guess who the author was. Not only did he sign the preface ("Advertisement") with the initials "I.N."; he also identified himself by stating that "Part of the ensuing Discourse about Light was written at the desire of some Gentlemen of the Royal Society, in the Year 1675" and was "sent to their Secretary, and read at their Meetings."
9. Newton might have conceived of omitting his name from this imperfect work, written in the vernacular language, from the prior example of Christiaan Huygens. Huygens, like Newton, considered his own treatise on light to be imperfect, and so published it in French as *Traité de la lumière* (1690), rather than in the universal language of scholarship and science, Latin. Huygens's work, like Newton's, did not bear the author's name on the title page, which merely states that the treatise is by "C. H. D. Z." There are some copies with the author's name spelled out on a special title page as "Christiaan Huygens Seigneur De Zuylichem."

10. The Latin edition (1706) contained seven new Queries, numbered 17–23; in the second English edition (1717/1718) these were rendered in English and revised and given the numbers 25–31. New Queries 17–24 were introduced into the second English edition. Some of the Queries at the end of the sequence were actually essays (or mini-treatises), occupying from two pages (Query 26) to four (Query 29) or nine (Query 28) pages and, in one case (Query 31), as much as thirty pages in print.
11. Preceding Query 1 of book 3; p. 339 of Dover edition.
12. Desaguliers said that some readers may have been led to think of the Queries as mere "conjectures," but Newtonians (including himself and Stephen Hales) "look upon them as we do in the rest of his works." Hales, according to Desaguliers, had "made a great many" experiments which "cleared up many of [Newton's] hints, and showed that Sir Isaac Newton had made no rash assertions." The Queries, Desaguliers concluded, "upon examination, appear to be true." See Cohen, *Franklin and Newton,* pp. 249–250, 254–255.
13. A bibliographical description of the editions of Franklin's book appears in section 4.1 of I. B. Cohen, *Benjamin Franklin's Experiments: A New Edition of Benjamin Franklin's Experiments and Observations on Electricity, Edited, with a Critical and Historical Introduction* (Cambridge, Mass.: Harvard University Press, 1941). See Chapter 1, Note 1.
14. Joseph Priestley, *The History and Present State of Electricity,* 3d ed. (London: C. Bathurst, 1755); facsimile reprint, with an introduction by Robert E. Schofield (New York: Johnson Reprint Corporation, 1966), vol. 1, p. 193.
15. Hélène Metzger, *Newton, Stahl, Boerhaave et la doctrine chimique* (Paris: Librairie Félix Alcan, 1930).
16. Pierre Duhem, *Le mixte et la combinaison chimique* (Paris: Gauthier-Villars, 1902).
17. See Note 13.
18. Later in our scholarly careers, both Henry Guerlac and I produced editions of the *Opticks* and we both wrote on Stephen Hales. I was responsible for the Dover reprint (see Note 1), of which the most significant novelty was the analytical table of contents. Guerlac worked for a number of years on a true scholarly edition, which was all but completed at the time of his death in 1984.
19. The "merely mathematical" was evidently a reference to the two mathematical tracts (see Note 6).
20. J. T. Desaguliers, *A Course of Experimental Philosophy,* 3d ed. (London: A. Millar, 1763), p. viii.
21. For the different editions of Algarotti's book, see Peter and Ruth Wallis, *Newton and Newtoniana, 1672–1975: A Bibliography* (London: Dawson, 1977).
22. Quoted in Metzger, p. 191.
23. Stephen Hales, *Vegetable Staticks* (London: W. & J. Innys and T. Wood-

ward, 1727); reprint with an introduction by Micheal A. Hoskin (London: Macdonald; New York: American Elsevier, 1969).

24. Stephen Hales, *Haemastaticks* (London: W. Innys & R. Manby and T. Woodward, 1733); facsimile reprint (Birmingham, Ala.: Classics of Cardiology Library, 1987).
25. See Cohen, *Franklin and Newton,* chap. 7, section 6.
26. Ibid., pp. 179–182.
27. See Note 7.
28. Cohen, *Franklin and Newton,* chap. 7.
29. Desaguliers, *Course of Experimental Philosophy,* vol. 1, p. xvi.
30. Ibid., p. 17; see *Franklin and Newton,* p. 251.
31. See Chapter 4.
32. Vol. 1, chap. 1.
33. On Desaguliers and Franklin, see Cohen, *Franklin and Newton,* pp. 243–246.
34. See "Supplement: The Franklin Stove."
35. There is evidence aplenty that Franklin had read Newton's *Opticks.* See Cohen, *Franklin and Newton,* chap. 7.
36. Humphry Davy, "Agricultural Lectures," part 2; in *Collected Works* (London: Smith, Elder, 1840), vol. 8, p. 264.
37. Sent to London in a letter addressed to Dr. John Mitchel, F.R.S. See Cohen, *Franklin's Experiments,* p. 201.
38. These postulates appear at the head of the article sent to Mitchel.
39. Letter 2 of the later editions of Franklin's book. See Cohen, *Franklin's Experiments,* pp. 171, 174.
40. Ibid., p. 171.
41. See Chapter 6.
42. Cohen, *Franklin's Experiments,* p. 172.
43. Ibid., p. 173n.
44. Ibid., p. 187; letter to Peter Collinson, 29 Apr. 1749, containing "Farther Experiments and Observations in Electricity, 1748."
45. See John L. Heilbron, *Electricity in the 17th and 18th Centuries* (Berkeley: University of California Press, 1979).
46. Cohen, *Franklin's Experiments,* p. 189, section 8.
47. Ibid., p. 191, section 16.
48. Ibid., pp. 191–192, section 17. In these lengthy extracts I have altered the paragraphing, introducing a new paragraph where Franklin's text contains a series of dashes.
49. Ibid., pp. 192–193, section 18. Franklin observed that he later learned "that Mr. *Smeaton* was the first who made use of panes of glass for that purpose" (p. 193n).
50. Franklin used the term "battery" for a battery of condensers; we have shifted the application to a series of dry or wet voltaic cells.
51. Cohen, *Franklin's Experiments,* pp. 180–181.
52. On the air condenser and its role in the demise of the Franklinian doctrine

of electrical atmospheres, see Heilbron, *Electricity in the 17th and 18th Centuries*, pp. 387–390.

53. In "Farther Experiments and Observations in Electricity, 1748," sent to Collinson on 29 Apr. 1749, Franklin (*Franklin's Experiments*, p. 199, section 28) described an "experiment . . . which surprises us." This experiment showed "that bodies having less than the common quantity of electricity, repel each other, as well as those that have more." This topic was considered again in a paper dated 14 March 1755, "in pursuance of" experiments made by John Canton (*Franklin's Experiments*, p. 302, section III; see also p. 76, n. 35). This problem—why negatively charged bodies should repel one another—was resolved by Aepinus's revision of the Franklinian theory. This revision, however, was occasioned by more fundamental problems, for which see the introductory monograph by R. W. Home to P. J. Connor's translation of Aepinus's *Essay on the Theory of Electricity and Magnetism* (Princeton: Princeton University Press, 1979). See, further, Heilbron, *Electricity in the 17th and 18th Centuries.*
54. The difficulties with Franklin's theory of electrical atmospheres have been analyzed most perspicaciously in Roderick Weir Home, *The Effluvial Theory of Electricity* (New York: Arno Press, 1981; reprint of Home's doctoral dissertation, Indiana University, 1967), esp. pp. 191 ff. See also Home's "Franklin's Electrical Atmospheres," *British Journal for the History of Science*, 6 (1972): 131–151. See also the work by Heilbron cited in Note 45.
55. Translated from the Latin of Musschenbroek's letter to Franklin, Leyden, 15 April 1759; printed in Jared Sparks, ed *The Works of Benjamin Franklin*, vol. 7, pp. 186–187.
56. There were five editions in English, three in French (in two different translations), and one each in German, Italian, and Latin. See Note 13.
57. Extract from Franklin's research journal, 7 Nov. 1749, quoted in a letter to Dr. John Lining, 18 Mar. 1755; see Cohen, *Franklin's Experiments*, p. 334.
58. This proposed experiment was described in full by Franklin in his book on electricity; see Cohen, *Franklin's Experiments*, p. 222, section 21. See also Chapter 6.
59. See Cohen, *Franklin's Experiments*, pp. 103ff, 256ff. After Franklin had published his account of the sentry-box experiment, it occurred to him that he did not need to wait for the completion of the steeple on Christ's Church, where he had planned to perform the test. He saw that he could achieve the same effect with the lightning kite. Evidently, he performed the kite experiment after the original sentry-box experiment in France had been carried out but before word had reached him of the successful outcome. Incidentally, for those who imagine that the experiment of the lightning kite was fraught with danger, I can report that some dozens of experimenters in the succeeding decades performed this experiment with safety and success.
60. William Watson, in *Philosophical Transactions*, 48 (1753): 201–202.
61. Sparks, *Works*, vol. 5, p. 501.
62. See Cohen, *Franklin's Experiments*, pp. 129–130.

63. See Chapter 1, Note 24.
64. See Chapter 8.
65. It is reproduced as the frontispiece of Cohen, *Franklin's Experiments.*

3. *How Practical Was Franklin's Science?*

Originally published as "How Practical Was Benjamin Franklin's Science?" *Pennsylvania Magazine of History and Biography,* 69 (1945): 284–293.

1. Curtis P. Nettels, *The Roots of American Civilization: A History of American Colonial Life* (New York, 1938), p. 454.
2. Ibid., pp. 497–498.
3. Franklin to Mary Stevenson, 20 Sept. 1761, quoted from A. H. Smyth, ed., *The Writings of Benjamin Franklin* (New York, 1905–1907), vol. 4, pp. 111ff; this letter is quoted at greater length and discussed in Chapter 9.
4. See Chapter 9.
5. Merle Curti, *The Growth of American Thought* (New York and London, 1943), p. 98. This splendid book tends to be more moderate than many others; the quoted statement is modified by the assertion that despite the fact that Franklin's conception of science was governed by utilitarianism, "it did not exclude a larger vision."
6. Smyth, *Writings,* vol. 2, pp. 456ff, 460ff. The reference is to Bernard Frénicle de Bessy, *Traité des triangles rectangles en nombre* (Paris, 1676).
7. The various editions of this book are listed and collated in I. Bernard Cohen, ed., *Benjamin Franklin's Experiments: A New Edition of Franklin's Experiments and Observations on Electricity* (Cambridge, Mass., 1941), pp. 139ff.
8. Smyth, *Writings,* vol. 2, p. 460; on Collinson's relationship with Franklin, see chapter 5.
9. "Autobiography," in Smyth, *Writings,* vol. 1, p. 233; scholars interested in consulting the autobiography should be aware of Benjamin Franklin, *Autobiography: An Authoritative Text, Backgrounds, Criticism,* ed. J. A. Leo Lemay and P. M. Zall (New York: Norton, 1986).
10. Reprinted in Jared Sparks, ed., *The Works of Benjamin Franklin* (Philadelphia, rev. ed., n.d.), vol. 2, pp. 66ff.
11. "Autobiography," in Smyth, *Writings,* vol. 1, p. 374.
12. Curti, *Growth of American Thought,* p. 98.
13. Spence was identified in my original article as Dr. Adam Spencer. Subsequent research revealed that Dr. A. Spencer's name was Archibald. See J. A. Leo Lemay "Franklin's 'Dr. Spence': The Reverend Archibald Spencer (1698?–1760) M.D.," *Maryland Historical Magazine,* 59 (1964): 199–216. For a detailed discussion of Spencer's influence on Franklin, see Chapter 4.
14. See Cohen, *Franklin's Experiments* p. 170; see Chapter 1, Note 1, for a reference to a new edition of Franklin's "Experiments and Observations"; see also Chapter 5.
15. Cohen, *Franklin's Experiments,* pp. 171ff, 179ff.

16. Ibid., 63.
17. Ibid., 199–200.
18. The chronology of the kite has been worked out in Carl Van Doren, *Benjamin Franklin* (New York, 1938); see also Chapter 6.
19. The relation of Franklin's fame and reputation as a scientist to his political and diplomatic career is discussed in I. Bernard Cohen, "Benjamin Franklin Scientist and Citizen," *The American Scholar,* 12 (1943): 47ff.
20. *See* Lois Margaret MacLaurin, *Franklin's Vocabulary* (Garden City, N.Y., 1928).
21. Joseph Priestley, *The History and Present State of Electricity, with Original Experiments* (London, 1755), vol. 1, p. 193.
22. Cohen, *Franklin's Experiments,* p. 279.
23. *See* Van Doren, *Franklin,* p. 700; I. Bernard Cohen, "Benjamin Franklin and Aeronautics," *Journal of the Franklin Institute,* 232 (1941): 104; see also Chapter 12.
24. *See* Bence Jones, *Life and Letters of Faraday* (London, 1870), vol. 1, p. 218; for more on Faraday, Franklin, and the baby, see Chapter 12.

4. *The Mysterious "Dr. Spence"*

Originally published as "Benjamin Franklin and the Mysterious 'Dr. Spence': The Date and Source of Franklin's Interest in Electricity," *Journal of the Franklin Institute,* 23 (1943): 1–25. I gratefully acknowledge the kindness of many friends and fellow researchers who have contributed information, criticism, and suggestions which have gone into this article: Carl Van Doren, Barney Chesnick of the Library Company of Pennsylvania, Clifford K. Shipton, Henry Guerlac, Gertrude D. Hess of the American Philosophical Society Library, Perry Miller, Edward Potts Cheney, Frederick B. Tolles, and Carl Bridenbaugh and Jessica Bridenbaugh.

1. The extreme popularity of Franklin's book on electricity may most easily be demonstrated by the number of editions—five in English, three in French, one in German, one in Italian. This book, which in the eighteenth century was printed in ten editions in four languages, has been reprinted as: *Benjamin Franklin's Experiments: A New Edition of Franklin's Experiments and Observations on Electricity,* edited with an introduction by I. Bernard Cohen (Cambridge, Mass.: Harvard University Press, 1941). This—the eleventh edition of the book—was, oddly enough, the first to be printed as a separate book in America. A facsimile of the *editio princeps* with a critical and textual commentary and a historical introduction by I. B. Cohen is scheduled for publication by the Burndy Library in collaboration with the Dibner Library, National Museum of American History, Washington, D.C.
2. On Franklin's self-education in the sciences, see I. B. Cohen, *Franklin and Newton* (Philadelphia: American Philosophical Society, 1956; Cambridge, Mass.: Harvard University Press, 1966), chap. 7.
3. The *Gentleman's Magazine* published many articles on electricity. One of

these, of special importance in Franklin's introduction to the subject, is discussed in Chapter 5.

4. Albert Henry Smyth, ed., *The Writings of Benjamin Franklin* (New York: Macmillan, 1905–1907), vol. 1, p. 417.
5. Ibid., p. 374.
6. Peter Collinson played a very important part in Franklin's scientific career. See Chapter 5.
7. Franklin to Michael Collinson, 1768 or 1769. Smyth, *Writings,* vol. 5, pp. 185–186. The new edition of *The Papers of Benjamin Franklin* gives the date 1770.
8. Collinson's "account of the new German Experiments in Electricity" is discussed in Chapter 5.
9. See J. A. Leo Lemay, "Franklin's 'Dr. Spence': The Reverend Archibald Spencer (1698?–1760) M.D.," *Maryland Historical Magazine,* 59 (1964): 200–201.
10. *The Pennsylvania Magazine of History and Biography,* 1 (1877): 246.
11. Ibid., pp. 413–414.
12. John Smith, a Quaker intellectual, was very much interested in the progress of science. See Frederick B. Tolles, "A Literary Quaker: John Smith of Burlington and Philadelphia," *The Pennsylvania Magazine of History and Biography,* 65 (1941): 300–333.
13. The originals of these notes are in the Library Company of Philadelphia (Ridgway Library). They may be found in Smith MSS., vol. 5, pp. 254–255. They cover both sides of two sheets of paper. It is by no means certain whether these notes represent one lecture or several. Nor can one tell what order the sheets should follow. I have therefore indicated the page separation by means of asterisks.
14. Alexander Hamilton, "Hamilton's Itinerarium, Being a narrative of a Journey from Annapolis, Maryland, through Delaware, Pennsylvania, New York, New Jersey, Connecticut, Rhode Island, Massachusetts and New Hanpshire, from May to September, 1744," ed. Albert Bushnell Hart (Saint Louis, 1907), pp. 232–233.
15. It was to Mitchel, rather than to Collinson, that Franklin addressed his first paper on the electrical nature of lightning. See Cohen, *Benjamin Franklin's Experiments.* An account of Mitchel (under the alternative spelling "Mitchell") may be found in the *Dictionary of American Biography.*
16. James Alexander and Cadwallader Colden were amateur scientists who were friends and scientific correspondents of Franklin. Colden, lieutenant governor of New York, was particularly interested in gravitation and electricity. Accounts of both appear in the *Dictionary of American Biography;* an account of Cadwallader Colden, written by Brooke Hindle, appears in the *Dictionary of Scientific Biography.*
17. *The Letters and Papers of Cadwallader Colden,* vol. 3, Collections of the New-York Historical Society (1919), p. 46.
18. Ibid., vol. 8 (1934), p. 321.

19. Ibid., p. 338.
20. Carl Bridenbaugh and Jessica Bridenbaugh, *Rebels and Gentlemen: Philadelphia in the Age of Franklin* (New York: Reynal and Hitchcock, 1942), p. 269; here "Spence" was mistakenly identified as Dr. Adam Spencer.
21. See Raymond P. Stearns, *Science in the British Colonies of America* (Urbana, Ill.: University of Illinois Press, 1970).
22. Lemay, "Franklin's 'Dr. Spence,' " pp. 204, 212.
23. J. A. Leo Lemay, *Ebenezer Kinnersley, Franklin's Friend* (Philadelphia: University of Pennsylvania Press, 1964), p. 52.
24. This last bit of information was originally turned up by Carl Bridenbaugh, who very kindly made it available to me. Gratz Collection of the Historical Society of Pennsylvania, case 7, box 21.
25. Carl Van Doren, *Benjamin Franklin* (New York: Viking, 1938), p. 140.
26. Carl Van Doren originally called my attention to this fact, which appears in "Franklin's Draft Scheme of the Autobiography," in Smyth, *Writings,* vol. 1.
27. For the vagaries in the publication of Franklin's autobiography, see Max Farrand, "Benjamin Franklin's Memoirs," *Huntingon Library Bulletin,* 10 (October 1936): 49–78.
28. *Life, Journals and Correspondence of Rev. Manasseh Cutler,* ed. William Parker Cutler and Julia Perkins Cutler (Cincinnati, 1888), vol 1, p. 269. See also Van Doren, *Franklin,* p. 750.
29. Franklin's two letters concerning the circulation of the blood were written to Colden in 1745 and may be found in Smyth, *Writings,* vol. 2, pp. 284–289, 290–294.
30. A convenient summary of the history of electricity prior to Franklin's research may be found in J. L. Heilbron, *Electricity in the 17th and 18th Centuries: A Study of Early Modern Physics* (Berkeley: University of California Press, 1979).
31. Stephen Gray, "Several Experiments Concerning Electricity," *Philosophical Transactions of the Royal Society,* 37 (1731–32): 39.
32. N. V. de V. Heathcote, "Franklin's Introduction to Electricity," *Isis,* 46 (1955): 29–35.
33. Ibid., p. 32.
34. Ibid., p. 35.
35. John L. Heilbron, "Franklin, Haller, and Franklinist History," *Isis,* 68 (1977): 509–526; see also Heilbron, *Electricity in the 17th and 18th Centuries,* p. 324, n. 3, and Bernard S. Finn, "An Appraisal of the Origins of Franklin's Electrical Theory," *Isis,* 50 (1969): 363–369.
36. See Chapter 2.
37. See Cohen, *Benjamin Franklin's Experiments,* chap. 3.
38. The information concerning the life of Isaac Greenwood was given to me by Frederick G. Kilgour. Further details concerning his life may be found in Clifford K. Shipton, ed., *Sibley's Harvard Graduates,* vol. 6 (Boston: Massachusetts Historical Society, 1942). Information about the Library Company Miniutes was provided by Barney Chesnick.

39. Kinnersley's scientific achievements, his lectures, and his life are described in Cohen, *Benjamin Franklin's Experiments*. See also Joseph A. Leo Lemay, "Franklin and Kinnersley," *Isis,* 52 (1961): 575–581, and Lemay, *Ebenezer Kinnersley, Franklin's Friend* Philadelphia: University of Pennsylvania Press, 1964).
40. On the topic of the popularization of science and the careers of itinerant public lecturers on science, see William Northrop Morse, "Lectures on Electricity in Colonial Times," *New England Quarterly,* 7 (1934): 364–374, and Stearns, *Science in the British Colonies of America,* pp. 506–511. The latter is the most complete account of the subject of scientific lectures in colonial America.

    On science in colonial America, see Brooke Hindle, *The Pursuit of Science in Revolutionary America* (Chapel Hill: University of North Carolina Press, 1956). On instruments and instrument-makers, see Silvio A. Bedini, *Early American Scientific Instruments and their Makers* (Washington, D.C.: Smithsonian Institution, 1964); S. A. Bedini, *Thinkers and Tinkers: Early American Men of Science* (New York: Scribner's, 1975); I. B. Cohen, *Some Early Tools of American Science* (Cambridge, Mass.: Harvard University Press, 1950; New York: Russell & Russell, 1967).

5. Collinson's Gift and the New German Experiments

1. See Norman G. Brett-James, *The Life of Peter Collinson* (London: Dunstan & Co., 1926).
2. *Gentleman's Magazine,* 15 (1745): 193–197; *American Magazine and Historical Chronicle,* 2 (1745): 530–537.
3. I. B. Cohen, *Franklin and Newton* (Philadelphia: American Philosophical Society, 1956; Cambridge, Mass.: Harvard University Press, 1966; rev. rept., 1990), pp. 431–432.
4. John Heilbron, "Franklin, Haller, and Franklinist History," *Isis,* 68 (1977): 539–549. On the use of the concept of electrical atmospheres, see R. W. Home, *The Effluvial Theory of Electricity* (New York: Arno, 1981), and "Franklin's Electrical Atmospheres," *British Journal for the History of Science,* 6 (1972): 131–151.
5. Collinson to Colden, 30 March 1745, *The Letters and Papers of Cadwallader Colden,* vol. 3, *1743–1747* Collections of the New-York Historical Society, (1919), p. 110.
6. Heilbron, "Franklin, Haller, and Franklinist History."
7. See Cohen, *Franklin and Newton,* pp. 392, 394, 394n, 402, 433.
8. Collinson to Colden, 30 March 1745, *Letters and Papers of Cadwallader Colden,* vol. 3, p. 110.
9. William Watson, *Sequel* (London, 1746); also in *Philosophical Transactions of the Royal Society,* 45 (1747): 704–749. A reason that Franklin may not have referred in this letter (or epistolary report) to the article in the *Gentleman's Magazine* is that he would have known that Collinson was aware of it, especially if Collinson either had sent a copy of it with the glass

tube, or had referred to the article in his enclosed directions for making experiments. With respect to Watson's *Sequel,* however, the situation was somewhat different; Franklin did refer to this work by title and author.

10. See Note 4.
11. Heilbron, "Franklin, Haller, and Franklinist History."
12. These German experiments have been summarized by Lloyd Espenschied, "The Electrical Flare of the 1740s," *Electrical Engineering,* 74 (1955): 392–397.
13. From Franklin's first report to Collinson.
14. I have developed this theme in *The Newtonian Revolution: With Illustrations of the Transformation of Scientific Ideas* (Cambridge, England: Cambridge University Press, 1980), and in *Revolution in Science* (Cambridge, Mass.: Harvard University Press, 1985).

6. *The Kite, the Sentry Box, and the Lightning Rod*

Originally published as "The Two Hundredth Anniversary of Benjamin Franklin's Two Lightning Experiments and the Introduction of the Lightning Rod," *Proceedings of the American Philosophical Society,* 96 (1952): 331–366. This essay was written with the assistance of a grant from the Penrose Fund of the American Philosophical Society. The bicentenary of the performance of Franklin's 1752 experiments provided the occasion for this study of the sequence of events in Franklin's research on lightning. Other studies of these topics appeared in 1952, notably B. J. F. Schonland, "The Work of Benjamin Franklin on Thunderstorms and the Development of the Lightning Rod," *Journal of the Franklin Institute,* 253 (1952): 375–391; a companion piece was my own study of "Prejudice against the Introduction of Lightning Rods," which appears below as Chapter 8. A more recent study of this topic, by Bern Dibner, appears in R. H. Golde, ed., *Lightning* (London, New York, San Francisco: Academic Press, 1977), vol. 1, pp. 23–49. For a somewhat different version of these events, see Pierre Brunet, "Les premières recherches expérimentales sur la foudre et l'électricité atmosphérique," *Lychnos* (1946–47): 117–148, and "Les origines du paratonnerre (discussions et réalisations)," *Revue d'Histoire des Sciences et de leurs Applications,* 1 (1948): 213–253.

1. See I. Bernard Cohen, *Franklin and Newton* (Philadelphia: American Philosophical Society, 1956; Cambridge, Mass.: Harvard University Press, 1966). A preliminary report appeared in the American Philosophical Society *Year Book* for 1949 (1950), 240–243; see also Cohen, "Benjamin Franklin, an Experimental Newtonian Scientist," *Bulletin of the American Academy of Arts and Sciences,* 5 (1952), 2–6.
2. N. H. Black, *An Introductory Course in College Physics,* 3d ed. (New York: Macmillan, 1948), p. 388.
3. *Proceedings of the American Philosophical Society,* 91 (1947): 17. Note that Franklin did not get "sparks between a kite-string and a key," but rather between the key and his knuckle.

4. See I. Bernard Cohen, *Benjamin Franklin's Experiments: A New Edition of Franklin's Experiments and Observations on Electricity* (Cambridge, Mass.: Harvard University Press, 1941), chap. 3; see Note 1, Chapter 1 above. See also Norman G. Brett-James, *The Life of Peter Collinson* (London: The Friends' Bookshop, 1926); Chapter 5; and Note 50.
5. This letter is reproduced with the kind permission of the Royal Society of London.
6. The text of this letter in *Philosophical Transactions of the Royal Society*, 47 (1751–52): 565–566, is identical, word for word, with the manuscript copy and follows the same paragraphing. The only difference between the two is in punctuation and spelling, e.g., "Publick" (in the MS) for "public" (in the *Transactions*), "*Philadelphia-Experiment*" for "Philadelphia experiment," "Electrick" for "electric," "easie" for "easy," "Ribon" for "riband," "joyn" for "join," "soone" for "soon," "comeing" for "coming," and generally, "&" for "and." When printed in the *Pennsylvania Gazette*, the *Gentleman's Magazine*, *London Magazine*, and several editions of his book on electricity, the letter bore the date of 19 October 1752 rather than 1 October 1752, and the final paragraph of the letter was missing. See Note 50.
7. Marcus W. Jernegan, "Benjamin Franklin's 'Electrical Kite' and Lightning Rod," *New England Quarterly*, 1 (1928): 180–196. The value of Priestley's testimony had been recognized earlier by James Parton, *Life and Times of Benjamin Franklin* (New York: Mason Brothers, 1864), vol. 1, p. 295, who in turn had obtained his information from Sparks. The latter prefaced the section of his edition containing the electrical papers with a pair of long extracts from Priestley and Stuber; Jared Sparks, ed., *The Works of Benjamin Franklin*, rev. ed. (Philadelphia: Childs & Peterson, 1840), vol. 5, pp. 172–180.
8. Letter from Priestley to Rotheram, dated 14 February 1766, quoted in Jernegan, "Franklin's Electrical Kite," p. 187. In the preface to his history, Priestley wrote: "With gratitude I acknowledge my obligations to Dr. Watson, Dr. Franklin, and Mr. Canton, for the books, and other materials with which they have supplied me, and for the readiness with which they have given me any information in their power to procure." Joseph Priestley, *The History and Present State of Electricity, with Original Experiments*, 3d ed. (London: C. Bathurst and T. Lowndes, 1775), vol. 1, p. xi. See W. Cameron Walker, "The Beginnings of the Scientific Career of Joseph Priestley," *Isis*, 21 (1934): 81–97, esp. pp. 87, 89.
9. Carl Van Doren, ed., *Benjamin Franklin's Autobiographical Writings*, (New York: Viking, 1945), p. 76.
10. Priestley, *History*, vol. 1, pp. 216–217. This account is preceded by a résumé of Franklin's letter to Collinson: "To demonstrate, in the completest manner possible, the sameness of the electric fluid with the matter of lightning, Dr. Franklin, astonishing as it must have appeared, contrived actually to bring lightning from the heavens, by means of an electrical kite, which he raised when a storm of thunder was peceived to be coming on.

This kite had a pointed wire fixed upon it, by which it drew the lightning from the clouds. This lightning descended by the hempen string, and was received by a key tied to the extremity of it; that part of the string which was held in his hand being of silk, that the electric virtue might stop when it came to the key. He found that the string would conduct electricity even when nearly dry, but that when it was wet, it could conduct it quite freely; so that it would stream out plentifully from the key, at the approach of a person's finger. At this key he charged phials, and from electric fire thus obtained, he kindled spirits, and performed all other electrical experiments which are usually exhibited by an excited globe or tube."

11. For a discussion of what is believed actually to occur in such an experiment, see Notes 61–64 and 98.
12. The "Bowdoin MS" was discovered by the writer during the preparation of *Benjamin Franklin's Experiments* and is described on pp. 152–154 of that work. It consists of a manuscript copy of all of Franklin's earliest papers on electricity, copied by two amanuenses and corrected by Franklin, and sent by Franklin to Bowdoin in 1750. As such it may be considered to replace the original MS copies of the letters, which no longer exist. In a number of instances, the "Bowdoin MS" gives a complete text, names of addressees, and dates, which were not to be found in previous printed editions of Franklin's book on electricity or collected editions of his writings; the text in *Benjamin Franklin's Experiments* was annotated to include the corrections made necessary by the "Bowdoin MS." See, further, I. B. Cohen, "Some Problems in Relation to the Dates of Benjamin Franklin's First Letters on Electricity," *Library Bulletin of the American Philosophical Society,* 100 (1956): 537–534.
13. Cohen, *Benjamin Franklin's Experiments,* p. 222.
14. A description of the various editions of Franklin's book on electricity may be found ibid.
15. In his autobiography Franklin wrote that "A Copy of them [my electrical papers] happening to fall into the Hands of the Count de Buffon . . . he prevail'd with M. Dalibard to translate them into French, and they were printed at Paris." Max Farrand, ed., *Benjamin Franklin's Memoirs, Parallel Text Edition* (Berkeley and Los Angeles: University of California Press, 1949), pp. 384–385. Dalibard, in the "avertissement" to his translation, noted that Franklin "pria en même tems M. Collinson d'en envoyer un des premiers exemplaires à M. de Buffon [qui] m'a engagé à les faire paroître en François" (first French edition, p. 4; second French edition, pp. 2–3; see Note 14).

    The Abbé Nollet, in *Lettres sur l'électricité* (Paris: H. L. Guérin & L. F. Delatour, 1754), a book devoted largely to an attack on Franklin's experiments and theories, notes that soon after the appearance of the English edition of Franklin's book (1751)

    > Un particulier qui reçut cet Ouvrage à Paris, le traduisit en François pour son propre usage, dit-on, & sans avoir dessein de le faire imprimer; cela est d'autant plus vraisemblable que cette traduction est un peu négligée; dans bien des en-

droits on a peine à entendre l'Auteur, & l'on manqueroit plusieurs de ses expériences, si l'on n'avoit pas recours à l'Original Anglois.

Quoi qu'il en soit, cet Ouvrage traduit dans notre langue, tomba entre les mains de M. de Buffon, Intendant du Jardin du Roi . . . Ce Sçavant Académicien, goûtant beaucoup la doctrine de M. Franklin, & ayant répété & vérifié, à ce que l'on prétend, avec succès toutes les expériences que cet Auteur apporte en preuves, crut obliger sa Patrie, en faisant publier cette traduction; & comme il étoit livré à des occupatíons plus importantes, il en abandonna le soin à de un ses amis nommé M. Dalibard, qui y joignit de son chef une histoire abrégée de l'Electricité; & cela forme un petit volume *in*-12 . . . (vol. 1, p. 5)

16. *Philosophical Transactions of the Royal Society,* 47 (1751–52): 534–535.
17. Dalibard published this memoir in full in the second edition of his translation of Franklin's book (1756), vol. 2, pp. 67–125, and the major portion of it was included in the fourth and fifth English editions of his book (1769 and 1774); see Note 14 and Cohen, *Benjamin Franklin's Experiments,* pp. 257–262.
18. *Philosophical Transactions of the Royal Society,* 47 (1751–52): 535.
19. Accounts of these experiments were printed ibid., pp. 536–552 (experiments of Mazéas, Le Monnier, and Ludolf as reported by Euler), 557–558 (Nollet and Le Roy), 559 (Mylius and Ludolf), 567–570 (Watson, Canton, Wilson, and Bevis). These experiments are discussed in more detail in Chapters 7 and 8 below. Another who repeated the sentry-box experiment in France was Romas, who claimed that he had conceived this experiment long before Franklin, just as he claimed that he had anticipated the kite experiment.
20. Farrand, *Franklin's Memoirs,* p. 386.
21. A. H. Smyth, ed., *The Writings of Benjamin Franklin* (New York: Macmillan, 1907), vol. 5, p. 94.
22. A. L. Rotch, "Did Benjamin Franklin Fly His Electrical Kite before He Invented His Lightning Rod?" *Proceedings of the American Antiquarian Society,* 18 (1907): 115–123; "When Did Franklin Invent the Lightning Rod?" *Science,* 24 (1906): 374–376; "The Lightning Rod Coincided with Franklin's Kite Experiment," ibid., p. 780.
23. A. McAdie, "The Date of Franklin's Kite Experiment," *Proceedings of the American Antiquarian Society,* 34 (1925): 188–205.
24. See Note 7.
25. Carl Van Doren, *Benjamin Franklin* (New York: Viking, 1938), pp. 164–170.
26. See Note 10.
27. Priestley, *History,* vol. 1, p. 206.
28. Rotch relied chiefly on the testimony of Romas that Franklin knew of the Marly sentry-box experiment before he flew his kite; he concluded (correctly) that the news would probably not have reached Philadelphia in June 1752. Romas was in error (see Note 31). Rotch's chief aim in the article, by his own admission, was to make known Franklin's directions for installing lightning rods, as published in "Poor Richard's Almanack" for 1753,

which, he maintained, "seems to have escaped the notice of all Franklin's biographers." Had Rotch consulted the best biography then available (Parton, *Life and Times,* vol. 1, p. 297n), he would have found this passage about lightning rods reprinted *in extenso.*

29. McAdie, "Date of Franklin's Kite Experiment."
30. Compare his translation of Franklin's book (see Note 14), vol. 2, p. 99. On pp. 72–73, he wrote, "There was lacking to that ingenious physicist [Franklin] but one last proof in order to produce complete conviction that the matter of thunder is absolutely the same as that of electricity; not being apparently too ready to acquire this proof by himself, he has shown us the means of obtaining it."
31. On this point Jernegan appears to have followed Park Benjamin, *A History of Electricity (The Intellectual Rise in Electricity) from Antiquity to the Days of Benjamin Franklin* (New York: John Wiley, 1898; rept., New York: Arno Press, 1975), p. 589. Benjamin assumed that Franklin had heard about the French experiment before he flew his kite; that he had believed the rod of Dalibard was not high enough (it was far from being as high as Franklin had specified) to draw electrical fluid from the clouds themselves. "That sparks had been drawn from rods which ended in the air close to the earth's surface, and not within hundreds of feet of the clouds was not conclusive. This was the experiment in one sense, and yet, in another, it was not. It showed that the rods had become electrified—but not necessarily that the lightning had electrified them or had passed over them." Benjamin believed that the kite experiment was designed to answer the doubt so aroused, but there is no evidence whatever to support his view. We do know that Franklin believed that pointed conductors could "draw off" electrical fluid from charged bodies at great distances and, indeed, this was the basis of his theory of the action of lightning rods. If he had overestimated the height necessary for a rod to draw off the supposed electrical fire from the clouds, I cannot see why this should have necessitated another experiment. We know that by September 1752, when he had certainly learned about the French experiments, he tested the electrification of clouds by the use of a rod he erected on his house in Philadelphia.
32. That is, lightning rods in the sense of grounded conductors to prevent buildings from damage by a stroke of lightning.
33. Smyth, *Writings,* vol. 3, p. 98.
34. This letter from the *Gentleman's Magazine* is reprinted above Note 53.
35. Van Doren, *Benjamin Franklin,* p. 168.
36. From *London Magazine,* May 1752.
37. See Note 28.
38. Smyth, *Writings,* vol. 3, p. 124.

    Volume 47 of the *Philosophical Transactions* (1751–52), containing the Abbé Mazéas's letter, the accounts of the sentry-box experiments in France and in England and Germany, and also the letter about the kite, was printed in 1753. On 20 July 1753, Collinson wrote to Franklin about the antics of the Abbé Nollet, who had been conducting a campaign against Franklin and whose "base and juggling intention" had been exposed: "Now on reading this & the King of France's approbation as you will see in the Transactions, if the Furror Schould Rise again what will allay it—This will lead you to understand my former Paragraph . . ." (from a letter in the Library of the American Philosophical Society).

39. The original manuscript letter is in the Library of the American Philosophical Society. See Cohen, *Benjamin Franklin's Experiments,* p. 124. Also published in *The Papers of Benjamin Franklin,* vol. 4 (New Haven: Yale University Press, 1961), pp. 333–335; the editor notes that "21 March" is probably an error for 20 March.
40. Manuscript letter in the Library of the American Philosophical Society. A letter from Collinson to Franklin, dated 15 August 1752, contains the following passage: "Our papers are full of Electrical Experiments. Thou sees a Little Electrical Hint give at Philadelphia has stimulated all Europe. I have not yet got a French translation. Expect it soone." See *Papers,* vol. 4, pp. 341–342.
41. Page 23; see Note 14.
42. Smyth, *Writings,* vol. 3, pp. 269–272; see *Papers,* vol. 6, pp. 97–101.
43. This letter (20 June 1754) is apparently lost. In Sparks, *Works,* vol. 6, pp. 193–194, there is an English translation of a letter from Dalibard to Franklin dated 31 March 1754, beginning: "I received on the 15th of January last, your obliging letter of October 28th." This letter is given under the incorrect date of 1752 by Jernegan, "Franklin's Electrical Kite," p. 182, n. 5, and by Edwin J. Houston, "Franklin as a Man of Science and an Inventor," *Journal of the Franklin Institute,* 161 (1906): 286. It appears in *Papers,* vol. 5 (1962), pp. 253–254.
44. From the second edition of Dalibard's translation (Paris, 1756), vol. 2, pp. 307–319. The remainder of the letter is as in Smyth, save that two short paragraphs are missing. The first (p. 312) states that Franklin "will be very glad to learn about the experiments of M. le Roy on positive and negative electricity, when you will be able to communicate them to me." The second (p. 319) is the conclusion of the letter: "I feel that the natural history of M. de Buffon will give me much pleasure & will instruct me enormously [*infiniment*]. Assure him, I beg of you, of my respects as well as M. de Fontserrière, both of whom have given me tokens of their regard in your last letter. I am *&c.* B. Franklin."
45. See Cohen, *Benjamin Franklin's Experiments,* pp. 307–310. Smyth does not indicate that this letter is merely an extract.
46. Letter in the Library of the American Philosophical Society.
47. Smyth, *Writings,* vol. 3, p. 98. If Franklin had not even received word of the French translation from Dalibard by 14 September, I assume he would not have received news of the Marly experiment directly from him until some time later than mid-September 1752.
48. Ibid., p. 164.
49. Smyth, ibid., pp. 105–106 (following Sparks), printed this letter under the date 1 January 1753, as it is also printed in *The Letters and Papers of*

Cadwallader Colden, vol. 5 Collections of the New-York Historical Society, (1920), pp. 358–359. In modern style the date should obviously be 1 January 1754, since Franklin did not send Nollet's book to Colden until 25 October 1753. Furthermore, the supplement to Franklin's book on electricity was not yet available in America on 1 January 1753.

50. See also Nathan Goodman, *A Benjamin Franklin Reader* (New York: Thomas Y. Crowell Co., 1945), pp. 381–382; Frank Luther Mott and Chester E. Jorgenson, *Benjamin Franklin: Representative Selections, with Introduction, Bibliography, and Notes* (New York: American Book Company, 1936), pp. 223–224.

    The editors of *The Papers of Benjamin Franklin* (vol. 4, pp. 365ff) have studied the problems of the date and text of Franklin's letter concerning the kite. They conclude that the date of 1 October is an error and that the probable date is 19 or 21 or 31 October. They also explore the textual problems of the final sentence about the erection of lightning rods.

51. G. Hellmann, "Über Luftelektricität," *Neudrücke von Schriften und Karten über Meteorologie und Erdmagnetismus,* 11 (Berlin, 1898). This issue contains facsimiles of publications by Winckler, Franklin, Dalibard, and Lemonnier. Rotch (see Note 22) called attention to Hellmann's discovery of the discrepancy between the two versions of this letter.

52. The various editors of Franklin's writings based their versions on previous editions, hence ultimately on the first supplement to the first English edition of Franklin's book on electricity, *Supplemental experiments and observations on electricity, part II. Made at Philadelphia in America, by Benjamin Franklin, Esq.; and communicated in several letters to P. Collinson, Esq.; of London, F.R.S.* (London: E. Cave, 1753). (This is E1.2 in the classification given in Cohen, *Benjamin Franklin's Experiments,* p. 141.) The final letter of that supplement, suppressed in the fourth and fifth English editions (1769 and 1774, respectively), printed immediately following the kite letter, was addressed to Collinson and read: "As you tell me our friend *Cave* is about to add some later experiments to my pamphlet, with the *Errata,* I send a copy of a letter from Dr. *Colden,* which may help fill a few pages; also my kite experiment in the *Pennsylvania Gazette;* to which I have nothing new to add, except the following experiment towards discovering more of the qualities of the electric fluid . . ." See *Benjamin Franklin's Experiments,* p. 161, n. 29.

53. *Gentleman's Magazine,* 22 (May 1752): 229.

54. Lemonnier or Le Monnier made a series of experiments, some in concert with Mazéas, with insulated rods and found that the experiment worked equally well if the rod was horizontal rather than vertical. His most significant results dealt with the electrification of the rod in the absence of thunderclouds, a topic brought to a high state of investigation by the Abbé Beccaria. For some account of Lemonnier's research, see Notes 19 and 51.

55. *Pennsylvania Gazette,* 28 Sept. 1752.

56. See Note 13.

57. For further information on lightning rods, see Chapter 8.

58. This letter was published in the *Gentleman's Magazine,* 20 (May 1750); 208. See Jernegan, "Franklin's Electrical Kite," p. 189, n. 26.
59. The description of the lightning rod in Poor Richard for 1753 has been often reprinted. We may note that it advocated grounded, pointed, metallic conductors for buildings and ships: "A House thus furnished will not be damaged by Lighting, it being attracted by the Points, and passing thro the Metal into the Ground [or water, in the case of ships] without hurting any thing." The complete issue of Poor Richard for 1753 is reproduced in facsimile in Mott and Jorgenson, *Representative Selections,* pp. 225–260.
60. Franklin first applied the concept of electrostatic induction to the charging of a Leyden jar. This point is discussed at length in Cohen, *Franklin and Newton.*
61. B. J. F. Schonland, *The Flight of Thunderbolts* (Oxford: Clarendon Press, 1950), p. 17.
62. The previous paragraph and the following one are based on the lucid discussion given by Schonland.
63. Schonland, *Flight of Thunderbolts,* p. 17.
64. Ibid., p. 23.
65. See Chapter 8.
66. Cohen, *Benjamin Franklin's Experiments,* pp. 307–308.
67. The Academy was the forerunner of the present University of Pennsylvania.
68. Although the issue of Poor Richard containing the account of the lightning rod was dated 1753, the text of it must have been prepared earlier than 19 October 1752, since the *Gazette* for the latter date carried an advertisement that "Poor Richard's Almanack" was "on the press."
69. Jernegan, "Franklin's Electrical Kite."
70. Smyth, *Writings,* vol. 3, p. 149. The discovery made by Franklin was that clouds appear to be charged negatively more often than positively and that therefore, "for most part, in thunder-strokes, *'tis the earth that strikes into the clouds, and not the clouds that strike into the earth.*"
71. A similar device was described by Dalibard in the second French edition of his translation of Franklin's book on electricity, vol. 1, p. lxxviii (histoire abrégée) and vol. 2, pp. 130–132, as having been invented by him a few days after he had made public his account of the Marly experiment (12 May 1752) and placed into operation on the rod which he had erected for Buffon at the *Jardin du Roi.* The pointed insulated rod ended in a bell, like Franklin's, but the second bell was not attached to a grounded rod but was merely attached "à la muraille."
72. Smyth, *Writings,* vol. 5, p. 421.
73. Ibid.
74. In 1758, while Franklin was in London, he wrote a letter to his wife about the bells: "If the ringing of the Bells frightens you, tie a Piece of Wire from one Bell to the other, and that will conduct the lightning without ringing or snapping, but silently." Smyth, *Writings,* vol. 3, p. 441. This letter plainly indicates that Franklin knew that the lightning was conducted through his rod system into the ground by a series of spark discharges as well as the

"convection" set up by the clapper, a device similar to that introduced by Franklin to discharge the two coatings of a Leyden jar and to prove that the charges on the two coatings were equal in absolute magnitude though opposite in sign.

The subsequent history of this rod system is described in Thomas Harrison Montgomery, *A History of the University of Pennsylvania from Its Foundation to A.D. 1770* (Philadelphia: George W. Jacobs Co., 1900), p. 75, n. 18: "These earlier experiments of Franklin were carried on in the house built by John Wister, No. 141 (now 325) Market street in 1731. 'It was in this house that Dr. Franklin . . . erected his first [?] lightning rod, an hexagonal iron rod, still in our possession, connecting it with a bell which gave the alarm whenever the atmosphere was surcharged with electric fluid. The ringing of the bell so annoyed my grandmother that it was removed at her request.' *Memoir of Charles J. Wister,* by his son, 1866, vol. i, pp. 21, 23. John Wister's son, Daniel, who was born 4 February, 1738–39, was a pupil at the Academy 1752–1754, as was also his cousin Caspar in 1752." It would be interesting to know whether the family papers of the Wisters, or of other Academy pupils in 1752, contain any references to "points" or to the kite experiment.

75. I believe that this important item, not hitherto cited in discussions of early lightning rods, confirms my earlier statement that Franklin would not have erected a wholly ungrounded lightning rod on a public building in Philadelphia in the summer of 1752. The implication is plain that lightning, attracted by the rod, would follow the metal as far as possible, but would then have to complete its path by traveling through plaster or woodwork, which it would rend and split. The rod erected by Franklin on his house in September 1752 avoided this danger since the choice of a path through metal meant that it had only to cross a small air gap and would not thereby damage the house. Hence, if Franklin had erected lightning rods in Philadelphia in the early summer of 1752, as I believe he did, they must have been grounded.
76. A close reading of the reports in the *Philosophical Transactions* (see Note 19) and those in the *Gentleman's Magazine* and *London Magazine* indicates that in every case the argument reads: since lightning is electrical, rods of some sort (grounding is not specified) can protect buildings or ships.
77. Smyth, *Writings,* vol. 4. p. 146.
78. Ibid., p. 129; on the use of lightning rods, see Chapters 7 and 8.
79. In other words, I have accepted Priestley's date of June 1752 and have indicated that because Franklin heard of the French experiments later in the summer, we have evidence consistent with Priestley's other statement that Franklin flew his kite before he learned what had been done in Europe. Others, for example Rotch (see Note 22), for whatever reasons, have assumed that Franklin must have performed the kite experiment after hearing of the French experiment and have then shown that he could not have flown the kite as early as June.
80. McAdie, "The Date of Franklin's Kite Experiment." This is the final part of

McAdie's conclusion, of which the remainder has been quoted above Note 29.

81. Quoted by McAdie. This version has been corrected by comparison with *The Complete Works in Philosophy, Politics, and Morals, of the Late Dr. Benjamin Franklin,* 2d ed. (London: J. Johnson & Longmans, Hurst, Rees, Orme, and Brown, n.d.), vol. 1, pp. 107–109.

82. See Franklin's letter to Dr. John Mitchel of 29 April 1749 "containing observations and suppositions, towards forming a new hypothesis for explaining the several phaenomena of thunder-gusts," in Cohen, *Benjamin Franklin's Experiments,* p. 209, in which after noting that "As electrified clouds pass over a country, high hills and high trees, lofty towers, spires, masts of ships, chimneys, *&c.* as so many prominencies and points, draw the electrical fire, and the whole cloud discharges there," he concluded, "Dangerous, therefore, is it to take shelter under a tree, during a thunder-gust. It has been fatal to many, both men and beasts."

83. *Complete Works,* vol. 1, p. 98.

84. Sparks, *Works,* vol. 5, p. 173.

85. Parton, *Life and Times,* vol. 1, p. 295.

86. From a letter written by Eddy to McAdie, 15 Dec. 1923, quoted by McAdie, "The Date of Franklin's Kite Experiment," pp. 16–17. Eddy wrote: "Parton, Vol. 1, page 289, says (referring to the spring of 1752), 'nearly three years have rolled away since he had suggested *in his private diary* a mode of ascertaining whether lightning and electricity were really the same.' I do not know what Parton meant by 'private diary.' I think he must have been referring to the paper written by Franklin in 1749 and entitled 'opinions and conjectures . . .' " Despite a number of errors, Parton usually knew what he was talking about and referred to genuine items. The "private diary" in question was one from which Franklin took an extract which he included in a letter to Dr. Lining dated 18 March 1755, which read: "Nov. 7, 1749. Electrical fluid agrees with lightning in these particulars: 1. Giving light. 2. Colour of the light. 3. Crooked direction. 4. Swift motion. 5. Being conducted by metals. 6. Crack or noise in exploding. 7. Subsisting in water or ice. 8. Rending bodies it passes through. 9. Destroying animals 10. Melting metals. 11. Firing inflammable substances. 12. Sulphureous smell.—The electric fluid is attracted by points.—We do not know whether this property is in lightning.—But since they agree in all the particulars wherein we can already compare them, is it not probable they agree likewise in this? Let the experiment be made." See Cohen, *Benjamin Franklin's Experiments,* p. 334.

87. I have been unable to locate any biographical information concerning Stuber, save an occasional mention of his name.

88. For Kinnersley's career, see Cohen, *Benjamin Franklin's Experiments,* Appendix 1; J. A. Leo Lemay, *Ebenezer Kinnersley: Franklin's Friend* (Philadelphia: University of Pennsylvania Press, 1964).

89. Smyth, *Writings,* vol. 3, pp. 95–97.

90. Van Doren, *Benjamin Franklin,* p. 165.

91. "One Paper which I wrote for Mr. Kinnersley, on the Sameness of Lightning with Electricity, I sent to Dr. Mitchel, an Acquaintance of mine, and one of the Members also of that society [i.e., the Royal Society of London]; who wrote me word that it had been read but was laught at by the Connoisseurs." Farrand, *Franklin's Memoirs,* p. 382.
92. For example, Collinson wrote Franklin on 5 Feb. 1750, "Your very Curious Pieces relating to Electricity and Thunder-Gusts have been read before the Society & have been Deservedly admired not only for the Clear Intelligent Stile, but also for the Novelty of the Subjects." Cohen, *Benjamin Franklin's Experiments,* p. 80; see also pp. 82–84.
93. Chiefly the Rev. William Stukeley, F.R.S., who mentioned Franklin by name in a paper explaining how earthquakes and lightning have the same cause—the electric fluid. See Cohen, *Benjamin Franklin's Experiments;* see also Chapter 8.

    Stukeley entered the following comment in his diary, after hearing Franklin's letter of 1 October read at the Royal Society: "21 December 1752. At the Royal Society. Mr Franklin, of Philadelphia, sent a pretty account of his extracting fire from the clouds, as singular in the invention as less operose and costly than those of the French astronomers. He makes a cross of two bits of cedar wood, tyes a silk handkerchief to the points by its corners, sets up a small iron half a foot long on that point which is the head of the kite, applys tail and wings to it as usual, a bunch of ribbands is to be between the end of the string and your hand, and then you fly it as ordinary kites when a cloud passes by loaden with the electric fire, and then you thus draw it down." "The Family Memoirs of the Rev. William Stukeley, M.D. (2)," *Publications of the Surtees Society,* 80 (1885): 466–467.

    Another who advocated the lightning-earthquake connection was Stephen Hales, the father of plant physiology; see Chapter 7. This topic is also explored in Michael N. Shute, *Earthquakes and Early American Imagination: Decline and Renewal in Eighteenth-Century Puritan Culture* (unpublished doctoral dissertation, University of California, Berkeley, 1977). See also Charles Edwin Clark, "Science, Reason, and an Angry God: the Literature of an Earthquake," *New England Quarterly,* 38 (1965): 340–362; Raymond P. Stearns, *Science in the British Colonies of America* (Urbana, Chicago, London: University of Illinois Press, 1970), pp. 642–670 (esp. pp. 648–650); Brook Hindle, *The Pursuit of Science in Revolutionary America, 1735–1789* (Chapel Hill: University of North Carolina Press, 1956), pp. 94–96.
94. Cohen, *Benjamin Franklin's Experiments,* p. 114, written apropos of the antics of Nollet in attempting to prove that Franklin's experiments would not work.
95. *Philosophical Transactions of the Royal Society,* 47 (1751–52): 569.
96. Van Doren, *Benjamin Franklin,* p. 169.
97. Sparks, *Works,* vol. 5, pp. 375–377.
98. Two further objections of McAdie to Priestley's account and Franklin's may be mentioned briefly. One is that the style ("tenor") of the letter

"indicates not so much an experiment actually performed as one projected and the results anticipated." I suppose McAdie referred to the form of statements such as, "You will find it [the electric fire] stream out plentifully . . ." or "the phial may be charged . . . spirits may be kindled," etc., rather than "I found it . . ." or "a phial was charged . . . spirits were kindled," etc. But a similar style was used by Franklin in earlier communications, for example, that to Collinson of 1 Sept. 1747, in Cohen, *Benjamin Franklin's Experiments,* pp. 179ff.: Experiment 3, "If a cork suspended by a silk thread hang between these two wires, it will play incessantly from one to the other . . ." Experiment 4, "Place an electrised phial on wax . . ."; Experiment 6, "Place a man on a cake of wax, and present him the wire of the electrified phial to touch, you standing on the floor, and holding it in your hand. As often as he touches it, he will be electrified *plus;* and any one standing on the floor may draw a spark from him." At the end of his "Farther experiments and observations in electricity, 1748," Franklin referred to some experiments he and his friends planned to perform, and the style is quite marked: "Spirits, at the same time *are to be* fired . . . A turkey *is to be* killed . . ." (my italics).

McAdie also states, "Franklin's conception, or perhaps the interpretation put upon the experiment and generally accepted, was that a cloud was a reservoir of electricity and the kite string a conductor. On the contrary, it appears to have been purely induction, not conduction. Had the kite string been wet enough to act as a conductor, the fibres would not have stood out." Two objections are indicated. First, the cord available in Franklin's day was a better conductor than McAdie supposed. For example, Stephen Gray's famous experiments on conduction and insulation and on induction made use of "pack-thread" for the conductor and silk lines for the insulator. Second, such cord is a better conductor when damp or wet, so that the system of metal in the kite, cord, and key was a better conductor when the rain had dampened the cord. Even if the phenomenon were, as pointed out earlier, one of induction and not conduction, the somewhat conductive (slightly) damp cord would produce the effect of fibres sticking out at a certain stage. McAdie may have underestimated the conductive power of eighteenth-century cord (packthread).

99. See Jérémie J. B. Abria, "Rapport sur l'éloge de Romas," *Actes de l'Académie des Sciences, Belles-Lettres et Arts de Bordeaux,* 15 (1853): 441–446. The Academy has also published a volume containing [1] Table historique et méthodique (1712–1875), [2] Documents historiques (1711–1713), and [3] Catalogue des Manuscrits de l'ancienne Académie (1712–1793) (Bordeaux: G. Gounouilhou, 1879); see p. 39 (no. 150), p. 61 (no. 349, no. 350), p. 62 (no. 351, no. 353, no. 355), for descriptions of his unpublished communications; for some account of MSS, see p. 185, p. 244 (no. 119), p. 245 (no. 142), p. 249 (no. 300, no. 315, no. 316, no. 317), p. 250 (no. 328 through no. 341), p. 278 (no. 1056), p. 279 (no. 1064)—the MSS in this list deal with scientific subjects, chiefly meteorology and electricity; they were (in 1877) in the Bibliothèque de la Ville (Bordeaux). We

may note that two of the above MSS dealt with the trisection of the angle and perpetual motion. See also p. 331 (no. 1552), p. 333 (no. 1556); and Notes 100, 106, and 109 below.

100. Romas, *Mémoire sur les moyens de se garantir de la foudre dans les maisons* (Bordeaux: Bergeret, 1776), p. 7.
101. Ibid., p. 12. This letter is printed on pp. 105–106; it was apparently read at a meeting of the Academy on 17 July.
102. Ibid., pp. 109, 132–133.
103. Ibid., p. 110.
104. Ibid.; Romas was mistaken, since Franklin's letter about the kite was read at a meeting of the Royal Society on 21 Dec. 1753.
105. Ibid. p. 133.
106. Merget, "Etude sur les travaux de Romas," *Recueil des Actes de l'Académie des Sciences, Belles-Lettres et Arts de Bordeaux,* 15 (1853): 447–511. The italics in the above quotation are Merget's (p. 484).
107. I have searched diligently through the first three English editions and the French edition, without finding any such passage as that quoted above.
108. Romas, *Mémoire,* p. 117.
109. Romas implied that these were the two pieces published by the Académie des Sciences in the second volume of the "Mémoires des savants étrangers." If so, they were MS copies, since that volume was not published until 1755; furthermore, only one memoir by Romas appeared in the second volume, and the other one appeared in the fourth volume, but it was dated 1757. "Mémoire, où après avoir donné un moyen aisé pour élever fort haut, & à peu de frais, un corps électrisable isolé, on rapporte des observations frappantes, qui prouvent que plus le corps isolé est élevé au dessus de la terre, plus le feu de l'électricité est abondant," *Mémoires de mathématique et de physique, présentés à l'Académie Royale des Sciences,* (1755): 393–407; "Copie d'une lettre écrite à M. l'Abbé Nollet par M. de Romas [De Nérac le 26 Aôut 1757]," ibid., 4 (1763): 514–517.

    For an English account, based on the first item, see *Gentleman's Magazine,* 26 (1756): 378–380.
110. Romas, *Mémoire,* p. 118.
111. Ibid., p. 109.
112. Freely translated from the French; printed in Romas (Note 109, first item), p. 395n.
113. Cohen, *Benjamin Franklin's Experiments,* p. 334.
114. Romas, *Mémoire,* p. 149. See M. l'Abbé Bertholon, *De l'électricité des météores* (Paris: Croullebois, 1787), vol. 1, pp. 32–55; also Nollet, *Lettres sur l'électricité* (Paris: H. L. Guérin & L. F. Delatour, 1754), vol. 2, pp. 228–248, asserting Romas's claims in a letter addressed to him.
115. Merget, "Etude," pp. 490–491. Merget had reference to the following statement "from Priestley": "MM de Lor et d'Alibard, dit-il, firent également l'expérience du cerf-volant en Angleterre, l'année suivante (ce qui est complètement faux), et M. de Romas voulant s'assurer par lui-même de ce

qu'il entendait raconter à ce sujet, la répéta en France avec beaucoup plus d'appareil." My comments in Note 107 apply equally well here.

116. Farrand, *Franklin's Memoirs,* p. 384.
117. Smyth, *Writings,* vol. 6, pp. 28–29.
118. Ibid., pp. 141–143.
119. Ibid., vol. 1, p. 14.
120. Ibid.
121. Ibid., vol. 5, p. 422n.
122. Priestley, *History of Electricity,* vol. 1, p. 411.
123. Sparks, *Works,* vol. 5, pp. 370–371.
124. *Philosophical Transactions of the Royal Society,* 48 (1754): 758. See *Gentleman's Magazine,* 23 (1753): 431.
125. Giambatista Beccaria, *A treatise upon artificial electricity . . . to which is added an essay on the mild and slow electricity which prevails in the atmosphere during serene weather, translated from the original Italian* (London: J. Nourse, 1776), pp. 449–450. For Franklin's high opinion of Beccaria's work, see his letter to Dalibard (Note 44); see also J. L. Heilbron, *Electricity in the 17th and 18th Centuries* (Berkeley: University of California Press, 1979), pp. 362–372.
126. Priestley, *History of Electricity,* vol. 1, p. 396.
127. See Paul Fleury Mottelay, *Bibliographical History of Electricity and Magnetism* (London: Charles Griffin, 1922), p. 320, for a table of eighteenth-century experiments on atmospheric electricity.

7. *Father Diviš and the First European Lightning Rod (with Robert Schofield)*

Originally published as "Did Diviš Erect the First European Protective Lightning Rod, and Was His Invention Independent?" *Isis,* 43 (1952): 358–363. For more on Diviš, see Josef Smolka's article on Diviš in the *Dictionary of Scientific Biography;* see also Karel Hujer, "Father Procopius Diviš and His Lightning Conductor," *American Journal of Physics,* 22 (1954): 108–109. An admirable article, based on extensive use of manuscript sources, is Josef Smolka, "The Correspondence of Procopius Diviš with L. Euler and the Academy of Sciences in St. Petersburg," *Sborník pro Dějini Přirodnich Věd a Techniky* (Československá Akademie Ved), 8 (1963): 139–162 (in Czech, with Russian summary). He has summarized his researches in "B. Franklin, P. Diviš et la découverte du paratonnerre," *Proceedings of the 10th International Congress of the History of Science,* 1962 (Paris: Hermann, 1964), part 2, pp. 763–767. See also Smolka, "Prokop Diviš and his Place in the History of Atmospheric Electricity," *Acta Historiae Rerum Naturalium Necnon Technicarum,* 1 (1965): 149–169. Recent sources in Czech are Jan Jajek, "Poznnámky k životopisu Prokopa Diviš" (Notes on the biography of Prokop Diviš), *Dějiny Věd Tech,* 11 (1978): 159–167, and Josef Haubelt, *Zivot a dilo Václava Prokopa Diviše* (Life and works of V. P. Diviš) (Vyoske: Okresni Museum ve Vysokém Mytě, 1982).

1. Karel Černý, *Prokop Diviš český vynálezce hromosvodu* (Procopius Diviš, the Czech inventor of the lightning rod) (Znojmo, 1948). See *Isis,* 40 (1949): 138.
2. Karel Hujer, "Father Procopius Diviš—the European Franklin," *Isis,* 43 (1952): 351–357.
3. Hujer, ibid., nn. 5 and 6, refers to the histories of physics by Poggendorff and Heller. A note on Diviš, with bibliography, appears in Paul Fleury Mottelay, *Bibliographical History of Electricity and Magnetism* (London, 1922), p. 209. Some account of Diviš may be found in Brother Potamian and James J. Walsh, *Makers of Electricity* (New York, 1909), 108–114. Further information concerning Diviš, and lists of articles describing his life and work, may be found in Léon Goovaerts, *Ecrivains, artistes et savants de l'Ordre de Prémontré: Dictionnaire bio-bibliographique,* vol. 1 (Brussels, 1899), pp. 195–197 (on p. 196, it is stated that Franklin reinvented the rod in 1760, whereas he had actually invented it a decade earlier, and had already begun erecting lightning rods in Philadelphia in 1752, two years before Diviš's rod of 1754); also vol. 4 (Brussels, 1911), p. 56.
4. From Franklin's autobiography, apropos of the controversy initiated by the Abbé Nollet. See Franklin's remarks to Ingenhousz concerning the latter's polemic against Priestley, in A. H. Smyth, ed., *The Writings of Benjamin Franklin* (New York, 1905–1907), vol. 1, p. 14; vol. 6, pp. 28–29.
5. Hujer, "Father Diviš," p. 355.
6. See Chapter 8.
7. Hujer, "Father Diviš," p. 357.
8. See I. Bernard Cohen, *Benjamin Franklin's Experiments* (Cambridge, Mass.: 1941), Introduction, chaps. 3 and 4; also see Chapter 5 above.
9. *Mémoires pour l'histoire des sciences & des beaux arts, commencés d'être imprimés l'an 1701 á Trévoux* (Paris, 1752), article 60, pp. 1208–1226.
10. Cohen, *Benjamin Franklin's Experiments,* pp. 256–262. Dalibard printed his report in the second edition of his translation of Franklin's book (Paris, 1756).
11. Jacques de Romas, *Mémoire sur les moyens de se garantir de la foudre dans les maisons* (Bordeaux, 1776), p. 33. For a discussion of Romas's independent conception of the lightning kite experiment, see Chapter 6.
12. Eufrosina Dvoichenko-Markov, "Benjamin Franklin and Leo Tolstoy," *Proceedings of the American Philosophical Society,* 96 (1952): 119.
13. *Nova acta eruditorum* (Leipzig, 1755), pp. 117ff, a review of Winkler's *Programma* (1753), referring explicitly to Franklin's book.
14. Abbé Nollet, *Lettres sur l'électricité* (Paris, 1753). Another vehicle for the dissemination of Franklin's ideas was a favorable book, Giovanni Baptista Beccaria, *Dell' electricismo naturale ed artificiale* (Turin, 1753). For more on Beccaria, see the *Dictionary of Scientific Biography* article by John L. Heilbron, as well as Heilbron's *Electricity in the 17th and 18th Centuries* (Berkeley, 1979), chap. 15, "The Reception of Franklin's Views in Europe," where the context for Nollet's response to Franklin's ideas is also discussed.
15. 22 Dec. 1753, pp. 1380, 1381, 1382. That Franklin's name was known in

Germany is also indicated by a reference to him under the heading "De electricitatis phaenomenis novis" in an article by Matthew Gesner, "De electro veterum, praelectio habita," read on 3 Feb. 1753 at Göttingen and published in *Commentarii Societatis Regiae Scientiarum Gottingensis,* 3 (1753): 67–114, esp. 114.

16. G. W. Richmann, "De indice electricitatis et eius usu in definiendis artificialis et naturalis electricitatis phaenomenis," *Novi Commentarii Academiae Scientiarum Imperialis Petropolitanae,* 1752–1753 (St. Petersburg, 1758), vol. 4, pp. 301ff. In the preliminary portion of the volume, pp. 33–35, there is an account of Richmann's death.
17. John Henry Winkler, *Elements of Natural Philosophy Delineated,* vol. 2 (London, 1757), pp. 70–71. The work by Mylius that Winkler refers to is undoubtedly the periodical *Physikalische Belustigungen,* edited by Mylius and published at Berlin by C. F. Voss, 1751–1757. A copy is in the Rare Book Collection of the Library of Congress. An "Inaugural-dissertation" on Mylius appeared in 1912, followed by another in 1914: Erwin Thyssen, *Christlob Mylius, sein Leben und Wirken* (Marburg, 1912); Rudolf Trillmich, *Christlob Mylius: Ein Beitrag zum Verständnis seines Lebens und seiner Schriften* (Halle a. S., 1914).
18. To be sure, Franklin's name was not always associated with the kite experiment, which he designed and executed in the summer of 1752, after Dalibard had performed Franklin's sentry-box experiment at Marly but apparently before Franklin had learned of the latter (see Chapter 6). Thus Johann Albrecht Euler (son of the famous mathematician) wrote a paper on the dynamics of kites in which he indicated that kites were not only children's toys but had been used by the celebrated Romas in his electrical experiments: "Des cerfs-volans," *Histoire de l'Académie Royale des Sciences et Belles Lettres,* 1756 (Berlin, 1758), 12: 322. Need I point out that, in any case, a discovery is not independent merely because the name of the person from whom ideas have come may be unknown?
19. *Philosophical Transactions of the Royal Society,* 47 (1751–52): 567–570.
20. *Letters to a German Princess,* ed. David Brewster, vol. 2 (New York, 1833), pp. 132–33.

Although Hujer states that "the problem of atmospheric electricity and of the lightning rod was a very remote subject to Euler," both of these topics are discussed at length in Letters to a German Princess, containing Euler's discussion about Diviš. Hujer also claims, "From various indications, we realize that Euler was also unaware of Franklin's existence." I cannot subscribe to this view in the light of the many discussions of Franklin's electrical ideas in the Berlin Academy in the early fifties. Thus Franklin's name figures prominently in a piece by Aepinus, "Mémoire concernant quelques nouvelles expériences électriques remarquables," *Histoire de l'Académie Royale des Sciences et Belles Lettres,* 1756 (Berlin, 1758), 12: 105–121, which concludes with the statement that these new experiments "sont extrémement favorables aux notions que Mr. Francklin a donnés de l'Electricité." Furthermore, Franklin's ideas are discussed (with continual mention of his name) by Euler's son in "Recherches sur la cause physique de l'électricité," ibid., 1757 (Berlin, 1759), 13: 125–159.

21. For the best account of the development of Franklin's views on the action of lightning rods, see B. F. J. Schonland, *The Flight of Thunderbolts,* (Oxford, 1950).
22. Franklin's note is based on a report in the July 1752 issue of the *London Magazine.*
23. These two reports are discussed in full in Chapter 6. While the Brussels rod may have been a test rod, and therefore insulated, the Paris rods seem to have been grounded.
24. *Gentleman's Magazine,* 20 (May 1750): 208.
25. I have long been certain that Franklin had erected a lightning rod on his own house in September 1752, but it now appears likely that he had previously erected two rods in Philadelphia—in June or July 1752. See Chapter 6.

8. *Prejudice against the Introduction of Lightning Rods*

Originally published in *Journal of the Franklin Institute,* 253 (1952): 393–440.

1. Franklin published an extract from this letter in the fourth and fifth editions of his book on electricity (London: 1769, 1774). See I. Bernard Cohen, ed., *Benjamin Franklin's Experiments: A New Edition of Franklin's Experiments and Observations on Electricity* (Cambridge, Mass.: Harvard University Press, 1941), p. 393; see also Chapter 1, Note 1.
2. Franklin's reply to Winthrop may be found in Cohen, *Benjamin Franklin's Experiments,* pp. 393–398.
3. According to Andrew Dickson White, *A History of the Warfare of Science with Theology in Christendom* (New York and London: Appleton, 1928; preface dated 1894), vol. 1, p. 344: "the means of baffling the powers of the air which came to be most widely used was the ringing of consecrated church bells. This usage had begun in the time of Charlemagne . . ."
4. Ibid., p. 347. When Franklin published his letter to Winthrop in the fourth edition of his book on electricity, he included a portion of the *oraisons* used "suivant le rituel de Paris, lorsqu'on bénit des cloches," which he had copied from an article of the Abbé Nollet, "Mémoire sur les effets du tonnerre comparés à ceux de l'électricité; avec quelques considérations sur les moyens de se garantir des premiers," *Histoire de l'Académie Royale des Sciences* (1764), "Mémoires," pp. 408–451.
5. White, *Warfare of Science with Theology,* vol. 1, p. 345. A sixteenth-century account of a bell consecration relates how the Bishop "sayde certen Psalmes, [and afterwards] he consecrateth water and salte, and mingleth them together, wherwith he washeth the belle diligently both within and without, after wypeth it drie, and with holy oyle draweth in it the signe of the crosse, and prayeth God, that whan they shall rynge or sounde that bell, all the disceiptes of the devyll may vanyshe away, hayle, thondrying, light-

ening, wyndes, and tempestes, and all untemperate weathers may be aswaged" (p. 346).

6. See I. Bernard Cohen, ed., *Benjamin Franklin: A Letter on Lightning Rods* (Cambridge, Mass., 1942).
7. White, *Warfare of Science with Theology*, pp. 347–349, offers evidence concerning the continuance in Protestant Germany of the belief that church bells might drive away tempests. They were rung for this purpose at least once in the twentieth century in England. Although the custom existed in Catholic countries in the eighteenth century, and in some even in the nineteenth, many officials of the eighteenth-century Catholic church—including Pope Benedict XIV—advocated the use of lightning rods.

   In Hasting's *Encyclopedia of Religion and Ethics* (New York: Scribner's, 1928), vol. 6, p. 315, it is postulated that at first the bells had been rung on the occasion of a violent storm in order to call the people together to pray for their mutual protection; then, over the course of time, the original purpose was lost and the bells were rung during storms without any assembling of the populace for prayer; the Devil, in other words, became vulnerable to the bells themselves and prayers were no longer needed to combat his storms. This idea does not seem well founded. It supposes that the Devil underwent a conditioning, like Pavlov's dogs, to the ringing of bells. Nor does it account for the prevalence of the custom of bell ringing in storms among primitive people, nor of the custom in modern Europe of firing cannons to dissipate lightning as well as ringing bells. More likely, the origin of the practice has traces of "sympathetic magic" in that storms, which are noisy disturbances in the atmosphere (produced by demons or the "powers of the air"), are supposed to be counteracted by a ritual producing a similar noisy disturbance.
8. Quoted by Franklin in his letter to Winthrop, from the article of Nollet (see Note 4). The original account of Deslandes is summarized in *Histoire de l'Académie Royale des Sciences* (1719), "Histoire," pp. 21–22.

   According to Deslandes, the people believed that the reason bell-ringing churches were struck was that the day was Holy Friday, when it is forbidden to ring church bells. The Rt. Rev. Monsignor John Walsh, in *The Mass and Vestments of the Catholic Church* (New York: Benziger Brothers, 1916), p. 339, states that according to Pope Benedict XIV, the bells are stilled from the "Gloria" in the Mass of Holy Thursday until the "Gloria" on Holy Saturday "because bells typify the preachers of the word of God and all preaching was silenced during the trial and passion of Our Lord."
9. Bacon had noted that "it is believed by some, that great ringing of bells in populous cities hath chased away thunder, and also dissipated pestilent air: all of which may be also from the concussion of the air, and not from the sound." See "Natural History," Cent. 2, 127, *Works of Francis Bacon*, ed. James Spedding, Robert Leslie Ellis, and Douglas Denon Heath (Boston: Brown & Taggard, 1862), vol. 4, p. 240.

   In the *Encyclopédie* of Diderot and d'Alembert (under the heading "Tonnerre"), we find the statement that "thunder can be disrupted and diverted by the sound of several bells or the firing of a cannon; thereby a great agitation is excited in the air which disperses the parts of the thunder . . ." This was written by the eminent Dutch physicist Pieter van Musschenbroek; he warned that the bells should never be rung when the clouds were exactly overhead, since then there would be great danger. Musschenbroek repeated these ideas in his oft-reprinted survey of physics (e.g., the French edition, *Essai de physique,* trans. Pierre Massuet [Leyden: Samuel Luchtmans, 1751], vol. 2, p. 846: chap. 40, "Des météores ignés," section 1732); it is interesting to note that although Musschenbroek believed in the physical action of bells to counteract lightning, he indicated the incorrectness of the view that held thunder and lightning to be "un effet du Démon & autres malins Esprits."

10. Nollet's objections to lightning rods are described at further length later in this chapter. For more on Nollet, see J. L. Heilbron, *Electricity in the 17th and 18th Centuries* (Berkeley: University of California Press, 1979), pp. 352–362.
11. See B. F. J. Schonland, *The Flight of Thunderbolts* (Oxford: Clarendon Press, 1950), p. 25. For those interested in the whole subject of lightning and the history of our knowledge of this phenomenon, Schonland's book cannot be too highly recommended.
12. Joseph Priestley, *History and Present State of Electricity,* 3d ed. (London: C. Bathurst, 1775), vol. 1, p. 472.
13. See Charles Richard Weld, *A History of the Royal Society* (London: John W. Parker, 1848), vol. 2, chap. 4. The committee dodged the issue, which arose soon afterward, between Wilson and the others as to whether the rods on buildings should be blunt or pointed, or whether they should project above the building. See Chapter 6. For more on Franklin's conductor, see Paul A. Tunbridge, "Franklin's Pointed Lightning Conductor," *Notes and Records of the Royal Society,* 28 (1974): 207–219.
14. W. Snow Harris, *On the Nature of Thunderstorms; and on the Means of Protecting Buildings and Shipping against the Destructive Effects of Lightning* (London: John W. Parker, 1843), pp. v-vi.
15. Ibid., p. 162; White, *Warfare of Science with Theology,* vol. 1, pp. 367–368.
16. Santiago Ramón y Cajal, *Recollections of My Life,* trans. E. Horne Craigie (Philadelphia: American Philosophical Society, 1937; *Memoirs,* vol. 8), pp. 21–23. In a recent biography of Ramón y Cajal, the author apparently could not bring herself to believe that anyone would have rung bells to prevent the effects of lightning; after describing the effects of the storm, she writes: "Just at that moment someone pointed to a black figure caught in the railing of the church belfry, its head hanging over the wall. It was the parish priest, who had tried to warn the people of the danger by tolling the church bell." Dorothy F. Cannon, *Explorer of the Human Brain* (New York: Henry Schuman, 1949), pp. 7–8. One would think that the thunder itself would provide a better (i.e., louder) "warning" than the bells; further-

more, what would the people be expected to do about the thunderstorm once they had been warned?

17. A description of the Swabian bells written by Ernst Meier is given in Ernest Morris, *Legends o' the Bells* (London: Sampson, Low, Marston & Co., n.d.), p. 20.
18. Ibid.
19. Ibid., pp. 16–17; the quotation is given by Morris from *Blackwood's Magazine,* Nov. 1867, p. 543.
20. Quoted in Morris, *Legends,* p. 21. In J. R. Nichols, *Bells thro' the Ages* (London: Chapman and Hall, 1928), p. 27, this extract is attributed to E. Morris in *The Bell News and Ringer's Record,* 23 Oct. 1915, p. 63.
21. J. J. Raven, *The Bells of England* (London: Methuen, 1906), p. 280.
22. Charles Burney, *The Present State of Music in Germany, the Netherlands, and United Provinces* (London: T. Becket, J. Robson, G. Robinson, 1775), vol. 1, p. 183.
23. The complete issue of "Poor Richard" for 1753 has been reproduced in facsimile in Frank Luther Mott and Chester E. Jorgenson, eds., *Benjamin Franklin: Representative Selections* (New York: American Book Company, 1936), pp. 255–260; the description of the lightning rods appears on p. 257.
24. Pedantic scholars often complain about the name "Poor Richard's Almanack," since the title page was of the form: *"Poor Richard Improved: being an Almanack and Ephemeris . . . for the Year of our Lord 1753: being the First after Leap Year . . . By Richard Saunders, Philom.* Philadelphia: Printed and Sold by B. Franklin, and D. Hall"; the annual didn't have "Poor Richard's Almanack" on the title page. In the advertisement, however, Franklin called his own work "Poor Richard's Almanack" and surely his own taste in this matter may serve as our guide.
25. See Cohen, *Benjamin Franklin's Experiments,* pp. 201–211. In Albert Henry Smyth, ed., *The Writings of Benjamin Franklin,* vol. 3 (New York: Macmillan, 1907), pp. 411–423, this letter is printed (erroneously) as if addressed to Peter collinson; no date is given in the text, although in the table of contents the date is given as "1750 (?)."
26. The action of points "in *drawing off* and *throwing off* the electrical fire" had been described by Franklin in his first communication to Peter Collinson on the subject of electricity, the letter dated 25 May 1747; see Cohen, *Benjamin Franklin's Experiments,* p. 170. Thomas Hopkinson had found that a pointed, insulated, charged conductor will rapidly be discharged (or "throw off the electrical fire") while Franklin had found that a pointed conductor, placed with the point facing a charged, insulated conductor, will discharge the latter (or "draw off the electrical fire")—slowly if the pointed conductor is insulated, rapidly if the pointed conductor is grounded.
27. Ibid., pp. 212–236.
28. B. F. J. Schonland discusses this same problem in a particularly illuminating way, in "Lightning Protection—200th Anniversary of the 'Philadelphia

Experiment,' " *Journal of the Franklin Institute,* 253 (1952): 375–504, esp. section 11, p. 384.

29. See Cohen, *Benjamin Franklin's Experiments,* pp. 82, 85, 87–88. Collinson wrote Franklin in April 1750 that his "electrical papers . . . are now on the press"; the *Gentleman's Magazine* (published by E. Cave, who also published Franklin's book on electricity) for May 1750 noted that Franklin's "ingenious letters on this subject [electricity] will soon be published in a separate pamphlet"; in the December 1750 issue of the *Gentleman's Magazine,* Cave referred to "an account of some experiments made in America by Mr. B. Franklin (now in the press)." See also Chapter 6.
30. This is the statement made by Dalibard in the "avertissement" to his translation (Paris, 1752), p. 4. For an account of the editions of Franklin's book, see Cohen, *Benjamin Franklin's Experiments,* pp. 141–148.
31. From the contemporaneous translation by James Parson, *Philosophical Transactions of the Royal Society,* 47 (1751–52): 534–535. Further lightning experiments in France may be found described in the same issue, pp. 536–552. On p. 559, there is an account of a performance of this experiment in Berlin. A detailed account of the experiment set up by Dalibard was included by him in the second edition of his translation of Franklin's book (Paris, 1756), and it may also be found in Cohen, *Benjamin Franklin's Experiments,* pp. 257–262.
32. See *Philosophical Transactions of the Royal Society,* 47 (1751–52): 567–570: "several members of the Royal Society . . . did, upon the first advices from France, prepare and set up the necessary apparatus for this purpose, [but] we were defeated in our expectations, from the uncommon coolness and dampness of the air here, during the whole summer." There had been but one thunderstorm in London and it had been accompanied by rain, which wet the apparatus. Canton, Wilson, and Bevis had performed the experiment away from London, but the effects they had obtained were "trifling" when compared to those reported from Paris and Berlin; yet, since "they were made by persons worthy of credit, they tend to establish the authenticity of those transmitted from our correspondents." This account is reprinted in Cohen, *Benjamin Franklin's Experiments,* pp. 262–264.
33. The question of exactly when Franklin flew his kite is treated in detail in Chapter 6.
34. *Philosophical Transactions of the Royal Society,* 47 (1751–52): 565–567.
35. Thus the rod used in the sentry-box experiment was ungrounded; the rods described in the letter of 29 July 1750 to Collinson, and in the 1753 "Poor Richard," were grounded.
36. In 1843, Snow Harris believed that the fear of unnecessarily inviting a lightning stroke was the chief cause of prejudice against lightning rods; pages 176–186 of *On the Nature of Thunderstorms* are devoted to this question. We are told that "the Governour-general and Council of the Honourable the East India Company were led to order the lightning rods to be removed from their powder magazines and other public buildings, hav-

ing in the year 1838 come to the conclusion, from certain representations of their scientific officers, that lightning rods were attended by more danger than advantage; in the teeth of which conclusion, a magazine at Dum Dum, and a corning-house at Mazagon, not having lightning rods, were struck by lightning and blown up" (pp. 176–177). Harris also quoted, from a book published in 1829, "Science has every cause to dread the thunder rods of Franklin: they attract destruction, and houses are safer without any of them" (p. 177).

37. For the nature of this controversy see Cohen, *Benjamin Franklin's Experiments;* Schonland, *The Flight of Thunderbolts;* and Weld, *History of the Royal Society*. See also Heilbron, *Electricity in the 17th and 18th Centuries,* pp. 379–82; 462; Tunbridge discusses this matter in "Franklin's Pointed Lightning Conductor."
38. Schonland, *The Flight of Thunderbolts,* p. 31.
39. Ibid.
40. Cohen, *Benjamin Franklin's Experiments,* pp. 307ff.
41. Ibid., p. 308n.
42. Smyth, *Writings,* vol. 6, p. 107. The first lightning rod for protective purposes to be erected in England appears to have been the one on Watson's house in 1762.
43. Ibid., vol. 1, p. 103.
44. Cohen, *A Letter on Lightning Rods*. De Saussure had erected a rod on his house in Geneva in 1771, which had caused so much anxiety to his neighbors that he feared a riot. To quell their fears he published a pamphlet that was later "translated in Italian and reissued at Venice," according to Douglas W. Freshfield, *The Life of Horace Benedict de Saussure* (London: Edward Arnold, 1920), p. 130. In Freshfield's list of de Saussure's writings, this pamphlet is listed as: "Exposition abrégée de l'Utilité des Conducteurs Electriques. 4to, pp. 9. Genève, 1771." An extract from this pamphlet is published in Richard Anderson, *Lightning Conductors: Their History, Nature, and Mode of Application* (London: E. & F. N. Spon, 1880), pp. 43–44. De Saussure wrote to Franklin from Naples on 23 February 1773 that the pamphlet had been successful in allaying the fears of his neighbors and that he had "had the pleasure of watching the electricity from the clouds during the whole course of the last summer. Several persons even followed this example, and raised conductors either upon their houses or before them. M. de Voltaire was one of the first. He does the same justice to your theory that he did to that of the immortal Newton." Quoted from Smyth, *Writings,* vol. 1, p. 104.
45. Cohen, *Benjamin Franklin's Experiments,* pp. 388ff.
46. Published in 1764 (see Note 4); for more on Nollet see Note 10.
47. Georg Wilhelm Richmann was electrocuted in St. Petersburg while performing an experiment in order to determine the "amount" of electricity in a lightning discharge by the use of an ungrounded test rod. This was the first fatal accident to occur during an electrical experiment and it was widely discussed throughout the world. For a list of contemporary accounts

of Richmann's death, see P. Fleury Mottelay, *Bibliographical History of Electricity and Magnetism* (London: Charles Griffin & Co., 1922), pp. 204–205.

48. A second edition—*Lettres sur l'électricité* (Paris: H. L. Guérin & L. F. Delatour, 1754)—contained a second group of letters, "Dans lesquelles on soutient le principe des effluences & affluences simultanées, contre la doctrine de M. Franklin, & contre les nouvelles prétentions de ses partisans." The translations from this book are largely based on those made by Watson for his review in *Philosophical Transactions of the Royal Society,* 48 (1753): 201–213.
49. For an account of the Franklin-Nollet controversy, see Cohen, *Benjamin Franklin's Experiments,* pp. 74–75, 108, 113–118.
50. See Mottelay, *Bibliographical History,* p. 209; also Brother Potamian and James J. Walsh, *Makers of Electricity* (New York: Fordham University Press, 1909), pp. 108–114.
51. The work of Diviš has been often discussed. A survey of the literature concerning him and his work may be found in Karel Hujer, "Father Procopius Diviš—the European Franklin," *Isis,* 43 (1952): 351–357. For a discussion of Diviš's lightning rod and further references concerning Diviš, see Chapter 7.
52. Mottelay, *Bibliographical History,* p. 209.
53. See Franklin's letter to Dalibard of 29 June 1755, in Cohen, *Benjamin Franklin's Experiments,* pp. 307ff; also Smyth, *Writings,* vol. 1, pp. 22, 199, 210; vol. 3, p. 269; vol. 4, pp. 141, 146, 457n; vol. 5, p. 165n.
54. Charles Burney, *The Present State of Music in France and Italy* (London: T. Becket & Co., 1771), p. 72.
55. Laura Caterina Bassi was not the only woman appointed professor at Bologna at this time; another was the famous mathematician Maria Gaetana Agnesi (the curve known as the "witch of Agnesi" is named after her). Consent to both appointments was given by Pope Benedict XIV; see Ludwig, Freiherr von Pastor, *The History of the Popes,* trans. E. F. Peeler (London: Routledge and Kegan Paul, 1949 [1950]), vol. 35, p. 192. Laura Bassi taught philosophy to Luigi Galvani and influenced him profoundly; see Potamian and Walsh, *Makers of Electricity,* p. 154.
56. Burney, *Music in France and Italy,* p. 217.
57. For example, he established a chair of surgery in Bologna and he was generally interested in the sciences. See Pastor, *History of the Popes,* vol. 35, chap. 3.
58. See Emile de Heeckeren, ed., *Correspondance de Benoît XIV* (Paris: Librarie Plon, 1912), vol. 2 [1750–1756].
59. Mottelay, *Bibliographical History,* p. 204.
60. Publications of Veratti are listed or described in Mottelay, pp. 186, 204, 213, 264, 384; Giovanni Fantuzzi, *Notizie degli scrittori bolognesi* (Bologna: Nella Stamperia di S. Tommaso d'Aquino, 1794), vol. 9, p. 193 (containing a brief biographical notice); Pietro Riccardi, *Biblioteca matematica italiana,* part 1 (Modena: Società Tipografica Modenese,

1876), vol. 2, p. 594; Serafino Mazzetti, *Repertorio di tutti i professori antichi e moderni della famosa università, e del celebre istituto delle scienze di Bologna* (Bologna: Tipografia di San Tommaso d'Aquino, 1847), p. 318, section 3098; Anderson, *Lightning Conductors,* pp. xv, 233 [1755[2]], 234 [1757].

61. See Mottelay, *Bibliographical History,* pp. 253–254; also Potamian and Walsh, *Makers of Electricity.* The need for rods was strikingly demonstrated in 1767, seventeen years after the lightning experiments of Franklin, when the Church of San Nazaro in Brescia was struck; some 200,000 pounds of gunpowder had been stored in the vaults, with no protection from lightning, and we are told that in the explosion "one sixth of the entire city was destroyed, and over three thousand lives were lost." See White, *Warfare of Science with Theology,* vol. 1, p. 368.
62. Snow Harris, *The Nature of Thunderstorms,* p. 97. The Latin inscription from the basilica in Assisi was made available to me by Paul Pascal of the University of Washington.
63. About thirty years after the invention of the lightning rod (according to White, *Warfare of Science with Theology,* p. 367), it was reported in Germany in 1783 that during thirty-three years "nearly four hundred towers had been damaged [by lightning] and one hundred and twenty bell-ringers killed." Schonland (*The Flight of Thunderbolts,* p. 8) refers to an edict of 1786 issued by the Parlement of Paris "to make the custom illegal on account of the many deaths it caused to those pulling the ropes. How necessary this edict was can be judged from figures given in a book published in Munich in 1784 with the cautious title, *A proof that the ringing of bells during thunderstorms may be more dangerous than useful.* The author, Fischer, stated that in 33 years lightning had struck 386 church towers and killed 103 bell-ringers at the ropes. These figures are not surprising when the accounts of lightning striking and damaging church steeples and other elevated structures are examined, for such damage was very frequent." Apparently the book in question, which I have not seen, was written by Johann Nepomuk Fischer, professor of mathematics at the University of Ingolstaat (1781–1876) and chief astronomer at the Mannheim observatory. J. C. Poggendorff, *Biographisch-literarisches Handwörterbuch* (Leipzig: Johann Abrosius Barth, 1863), vol. 1, p. 751, lists this as "Beweiss, dass das Glockenläuten bei Gewittern mehr Schädlich als Nützlich sey usw., Münch., 1784."
64. See Note 48.
65. *The Letters and Papers of Cadwallader Colden,* vol. 4, *1748–1754,* Collections of The New-York Historical Society (1920), p. 382.
66. See Smyth, *Writings,* vol. 3, pp. 149–153, 193. We would alter Franklin's statement to read "the base" of "the lower part" of the clouds; see Schonland, *The Flight of Thunderbolts,* section 6, p. 380.
67. Extract from the MS "Junto Minute Book" in the American Philosophical Library. I gratefully acknowledge the assistance given me by Gertrude Hess of the American Philosophical Society Library, who brought to my atten-

tion this Junto discussion, and whose wealth of knowledge concerning every aspect of Franklin's career has helped me and others to answer many vexing questions.

68. See Cohen, *Benjamin Franklin's Experiments* for an account of the life and work of Kinnersley; see also J. P. Leo Lemay, *Ebenezer Kinnersley, Franklin's Friend* (Philadelphia: University of Pennsylvania Press, 1964).

69. The full text of the description of Kinnersley's lectures is reprinted in William Northrop Morse, "Lectures on Electricity in Colonial Times," *New England Quarterly*, 7 (1934): 363–374.

70. Kinnersley's Newport broadside is dated 6 March 1752, printed by James Franklin. Two copies of this broadside exist. One is in the Rider Collection in the Brown University Library at Providence, described in George P. Winship, *Rhode Island Imprints* (Providence: Preston and Rounds, 1914), p. 13; reproduced in full facsimile on the plate facing p. 14. A second copy, in the private collection of A. S. W. Rosenbach of Philadelphia, formed one of the items in an exhibition held at the Free Library of Philadelphia in 1938. It is described in the catalogue of that exhibition, *The All-Embracing Doctor Franklin* (Philadelphia: Free Library of Philadelphia, 1938).

71. Original in the Library of Congress. It was found among the papers of William Thornton by Chauncey Worthington Ford, who printed the text of it in "New Light on Franklin's Electrical Experiments," *The Nation*, 86 (1908): pp. 85–86.

72. A photostat of the announcement of Johnson's course was provided by the Virginia Historical Society; information about the dates was kindly supplied by Mrs. J. A. Johnston, Assistant Secretary of the Society.

73. Information kindly supplied by Ellen M. FitzSimons, Librarian, Charleston Library Society.

74. See Cohen, *Benjamin Franklin's Experiments*, pp. 72n, 111, 150, 161, 331, 339; also Franklin C. Bing, "John Lining, an Early American Scientist," *Scientific Monthly*, 26 (1928): 249–252, and the article in the *Dictionary of American Biography*.

75. See Bernard Faÿ, *Franklin, the Apostle of Modern Times* (Boston: Little, Brown, 1929), p. 227. The third volume of the original French edition has a valuable bibliography and complete annotation: *Benjamin Franklin: Bibliographie et étude sur les sources historiques relatives à sa vie* (Paris: Calmann-Lévy, 1931), esp. p. 45.

76. In addition to White, *Warfare of Science with Theology*, see Theodore Hornberger, "The Science of Thomas Prince," *New England Quarterly*, 9 (1936): 24–42; Eleanor Tilton, "Lighting-rods and the Earthquake of 1755," *New England Quarterly*, 13 (1940): 85–97; Brooke Hindle, *The Pursuit of Science in Revolutionary America, 1735–1789* (Chapel Hill: University of North Carolina Press, 1956), pp. 94–96; Charles Edwin Clarke, "Science, Reason, and an Angry God: The Literature of an Earthquake," *New England Quarterly*, 38 (1965): 340–362; Michael N. Shute, *Earthquakes and Early American Imagination: Decline and Renewal in Eigh-*

teenth-Century Puritan Culture (unpublished doctoral dissertation, University of California, Berkeley, 1977).

77. Thomas Prince, *Earthquakes the Works of God and Tokens of His Just Displeasure . . . on Occasion of the Late Dreadful Earthquake which happened on the 18th of Nov. 1755* (Boston: D. Fowle & Z. Fowle, 1755), "Appendix," p. 23. This work (Evans 7549) was in part a reprint of an earlier work, the first portion of whose title was identical (Boston: D. Henchman, 1727; Evans 2946).
78. See Alfred Owen Aldridge, "Benjamin Franklin and Jonathan Edwards on Lightning and Earthquakes," *Isis,* 41 (1950): 162–164. Aldridge shows that the major part of the article on earthquakes was a word-by-word transcription from Ephraim Chambers's *Cyclopaedia* (London, 1728).
79. See Cohen, *Benjamin Franklin's Experiments,* Introduction, chap. 2.
80. Stukeley's articles in the *Philosophical Transactions* were reprinted in London in 1750 under the title *The Philosophy of Earthquakes.* Stukeley presented a copy of the third edition (1768) to Franklin, with the inscription "To Benjamin Franklin Esq. Father of Electricity. The Author." This copy, at present in the Boston Public Library, is described in Catalogue no. 943 of Davis & Harvey (auctioneers of Philadelphia), compiled by Stan. V. Henkels, *The Extraordinary Library of Hon. Samuel W. Pennypacker, Governor of Pennsylvania . . .* (1905).
81. Cohen, *Benjamin Franklin's Experiments,* pp. 166–167.
82. Some information on this hostility, and also on the reception of Franklin's ideas by Stukeley, may be found in an entry in the latter's diary, under the date of Dec. 1752, giving a brief survey of the history of electricity from 1704 to 1752: "Nov. 9, 1749, Mr. Collinson gave into the Royal Society Mr. Franklin, of Philadelphia, his discourse on thunder, lightning, fireballs, aurora borealis, and the like meteorological phaenomena, which he judiciously solves from electricity. 21 Dec. following, Mr. Collinson gave in another paper from Mr. Franklin on electricity. All these were printed, and a copy sent to France, which has excited the French philosophers, under the personal inspection of the monarch, to try so many experiments proving Mr. Franklin's doctrines of the cause of lightning, thunder, &c., from electricity. 15 Mar., 1749–50, after 2 shocks of earthquake we felt at London, I gave in a paper tending to prove that earthquakes are the effect of an electrical vibration of the surface of the earth, which has since been admitted by the French philosophers, though they seem to attribute it to Dr. Stephen Hale[s]. My paper was postponed by reason of other discourses till the next week, 22 March, when it was read, but with great opposition. 29 Mar., 1750, I gave in a second paper of 16 quarto pages, on earthquakes, confirming my former sentiments. It was deferred to the Thursday following, 5 April, 1750, when it was read. After it Dr. Hale's *[sic]* paper was read . . . Dr. Hale[s] and Mr. Flamsted before him make an approach toward electricity being the cause of earthquakes, but do not directly attribute them to it." "The Family Memoirs of the Rev. William Stukeley," vol.

2, *Publications of the Surtees Society,* 76 (1883): 378–379. See Stuart Piggott, *William Stukeley, an Eighteenth-Century Antiquary* (Oxford: Clarendon Press, 1950).

83. Prince, *Earthquakes the Works of God,* p. 20.
84. John Winthrop, *A Lecture on Earthquakes; Read in the Chapel of Harvard College in Cambridge, N. E. November 26th 1755* (Boston: Edes & Gill, 1755).
85. Ibid., p. 33.
86. See I. Bernard Cohen, *Some Early Tools of American Science* (Cambridge, Mass.: Harvard University Press, 1950).
87. See Franklin Bowditch Dexter, ed., *Extracts from the Itineraries and Other Miscellanies of Ezra Stiles . . . with a Selection from his Correspondence* (New Haven: Yale University Press, 1916), p. 595–596.
88. Prince, *Earthquakes the Works of God,* p. 20.
89. See the letters of the Abbé Mazéas to Stephen Hales, *Philosophical Transactions of the Royal Society,* 47 (1751–52): 536; Priestley, *History of Electricity,* vol. 1, section 11, pp. 421ff.
90. White, *Warfare of Science with Theology,* vol. 1, p. 348.
91. On the doctrine of "divine providences" in New England theology, the interpretation of various types of natural events as signs and portents of God's will, see Perry Miller, *The New England Mind* (New York: Macmillan, 1939), esp. chap. 8, "Nature"; Clifford K. Shipton, "The New England Clergy of the 'Glacial Age,' " *Publications of the Colonial Society of Massachusetts,* 32 (1936): 24–54.
92. Winthrop wrote a reply to Prince's letter, issued as a pamphlet, *A Letter to the Publishers of the Boston Gazette, &c. Containing an Answer to the Rev. Mr. Prince's Letter inserted in said Gazette* (Boston, 1756).

    Josiah Quincy, *History of Harvard University* (Boston: Crosby, Nichols, Lee & Co., 1860), vol. 2, pp. 219–220, relates that in 1770, "religious scruples were again raised against these protecting instruments, by representing 'thunder and lightning to be "tokens of the Divine displeasure," and that it was a degree of impiety to endeavor to prevent them from doing their full execution.' Professor Winthrop again appeared in defence of the invention of Franklin, by publishing a dissertation, adapted to counteract these scruples, and showing, that 'Divine Providence did not govern the material world by immediate and extraordinary interpositions of power, but by stated general laws'; and that it is as much 'our duty to secure ourselves against the effects of lightning, as from those of rain, snow, or wind, by the means God has put into our hands.' " At this time, Winthrop wrote to Franklin (26 Oct. 1770): "I have on all occasions encouraged [your] lightning rods . . . in this country, and have the satisfaction to find that it has not been without effect. A little piece I inserted in our newspapers last summer induced the people of Waltham (a town a few miles from hence) to fix rods upon their steeple, which had just before been much shattered and set on fire by lightning. They are now becoming pretty common among us, and numbers of people seem convinced of their efficacy." *Proceedings of the Massachusetts Historical Society,* 15 (1876–77): 11–13.

93. This and the following quotations from Adams may be found in Zoltán Haraszti, "Young John Adams on Franklin's Iron Points," *Isis,* 41 (1950): 11–14. See Haraszti, *John Adams and the Prophets of Progress* (Cambridge, Mass.: Harvard University Press, 1952).
94. Anderson, *Lightning Conductors,* pp. 169–175.
95. Schonland, *The Flight of Thunderbolts,* pp. 101, 59, 99.
96. Personal communication.
97. The actual financial losses are even greater, since these figures refer only to structures. Because many lightning-caused fires are in farm buildings (barns, silos, granaries, and so on), there may also be a loss of livestock, equipment, grain, and supplies, and a consequent loss of income.
98. For an analysis of the Prometheus myth, see Sigmund Freud, *Civilization and Its Discontents,* trans. Joan Riviere (London: Hogarth Press, 1951), esp. pp. 50–51, and "The Acquisition of Power over Fire," in *Collected Papers,* ed. James Strachey (London: Hogarth Press, 1950), vol. 5, pp. 288–294; Karl Abraham, *Dreams and Myths,* trans. William A. White (New York: The Journal of Nervous and Mental Diseases Publishing Company, 1913); Ernest Jones, *Essays in Applied Psycho-Analysis* (London and Vienna: International Psycho-Analytical Press, 1920), esp. p. 286.
99. Nollet, *Lettres,* pp. 17–18.
100. See Karl Abraham, *Selected Papers,* trans. Douglas Bryan and Alix Strachey (London: Hogarth Press, 1927), p. 233.
101. Quoted in Benjamin, *History of Electricity,* p. 592.

9. *Heat and Color*

Originally published as "Franklin's Experiments on Heat Absorption as a Function of Color," *Isis,* 34 (1943): 404–407; and "Franklin, Boerhaave, Newton, Boyle, and the Absorption of Heat in Relation to Color," *Isis,* 46 (1955): 99–104. In the second part of this chapter, which is based on the latter article, capitalization in quoted passages has been modernized.

1. Mary Stevenson later became Mrs. Hewson. She is also known as "Polly." An account of Franklin's relationship with her may be found in Carl Van Doren, *Benjamin Franklin* (New York: Viking, 1938).
2. The fourth edition was printed in London in 1769, the fifth in London in 1774.
3. A. H. Smyth, ed., *The Writings of Benjamin Franklin* (New York: Macmillan, 1905–1907), vol. 4, pp. 111ff. Smyth's version is a transcript of the original letter, then in the possession of T. Hewson Bradord, a descendant of Mary Stevenson's husband, William Hewson.
4. I. Minis Hays, ed., *Calendar of the Papers of Benjamin Franklin in the Library of the American Philosophical Society* (Philadelphia: American Philosophical Society, 1908), vol. 4, p. 173.

5. An outstanding example of this practice is the copy of his letters on electricity which Franklin had made in 1750 for James Bowdoin. These are described in detail in I. Bernard Cohen, *Benjamin Franklin's Experiments* (Cambridge, Mass.: Harvard University Press, 1941), chap. 4.
6. Published from the original in the American Philosophical Society Library, *Franklin Papers,* vol. 49, p. 37.
7. See Stanley Bloore, "Joseph Breintnall, First Secretary of the Library Company," *Pennsylvania Magazine of History and Biography,* 59 (1935): 42–56. The various aspects of Franklin's relations with Breintnal are recorded in Franklin's Autobiography. See Van Doren, *Benjamin Franklin.*
8. *Philosophical Transactions,* 41 (1739–40): 359. A second letter concerning a rattlesnake bite was read at a meeting of the Royal Society after Breintnal's death and is published ibid., 54 (1746): 147–150.
9. Published from the original in the Historical Society of Pennsylvania, *Logan Papers,* vol. 10, p. 100.
10. The English calendar was reformed in 1752, at which time the older Julian calendar was abandoned and the Gregorian calendar accepted. At this time, too, the beginning of the year was set at January 1, rather than at March 25, which had previously marked the beginning of the year. During the century or so prior to this reform, most writers used the form "1751/2" or 1751, 2" for dates between January 1, and March 25, in order to avoid confusion.
11. Bloore, "Joseph Breintnall."
12. Jared Sparks, ed., *Works of Benjamin Franklin,* rev. ed. (Philadelphia: Childs & Peterson, n.d.), vol. 2, p. 552.
13. Ibid., footnote on pp. 9–10.
14. Cohen, *Benjamin Franklin's Experiments,* p. 149.
15. *A new method of chemistry; including the history, theory, and practice of the art: translated from the original Latin of Dr. Boerhaave's Elementa chemiae, as published by himself* (London: T. Longman, 1741), vol. 1, pp. 262–263.
16. See Tenney L. Davis, "The Vicissitudes of Boerhaave's Textbook of Chemistry," *Isis,* 10 (1928): 33–46; Menno Hertzberger, *Short-Title Catalogue of Books Written and Edited by Hermann Boerhaave,* compiled with the assistance of E. J. van der Linden, preface by J. G. de Lint (Amsterdam: Hertzberger, 1927).

    The "first edition" of the Shaw and Chambers English version of Boerhaave's *Chemistry* was not based exclusively on the spurious edition printed in 1724 in "Paris" (actually Leiden), but also on various manuscript versions. Thus William Burton, writing about Boerhaave, thought this English version to be more reliable for Boerhaave's early views than the Latin one. I am greatly indebted to Henry Guerlac for clarification of the issues concerning the editions and translations of Boerhaave's *Chemistry.*
17. The account of this subject in the spurious edition reads: "Thus if you hang up several different coloured clothes, in a dark place; some white, others red, and others black; and place your self at a little distance: you will perceive nothing at all where the black are; where the white are, there will

something appear; so altho where the red are, tho' less than where the white. Not that there is more fire in one than another; but one reflects more or fewer rays than another; so that the circumstances of light do not depend on fire, so much as on the surface of the body that reflects it." Boerhaave, *A New Method of Chemistry;* trans. P. Shaw and E. Chambers, 2 vols. (London: J. Osborn & T. Longman, 1727), vol. 1 pp. 229–230.

18. Richard Waller, trans., *Essayes of Natural Experiments made in the Academie del Cimento . . .* (London, 1684; facsimile edition, with a new introduction by A. Rupert Hall, New York: Johnson Reprint Corporation, 1964).
19. See Hélène Metzger, *Newton, Stahl, Boerhaave et la doctrine chimique* (Paris: Félix Alcan, 1930). Although the text of Boerhaave's book abounds in expressions of admiration for the British school, citing Hooke, Halley, Bacon, and Boyle, the most flattering references to these authors, including long extracts, are to be found in the footnotes written by Shaw. Despite Boerhaave's admiration for the British school, he did not follow them in their "kinetic" concept of heat, but advocated a "fluid" concept. This point is explored in greater detail in I. B. Cohen, *Franklin and Newton* (Philadelphia: American Philosophical Society, 1956; Cambridge, Mass.: Harvard University Press, 1966; rev. rept., Harvard University Press, 1989).
20. The quotations from Newton's *Opticks* are taken from the fourth edition, reprinted by Dover Publications, 1952.
21. See J. F. Fulton, *A Bibliography of the Honourable Robert Boyle, Fellow of the Royal Society,* (Oxford: Oxford Bibliographical Society, 1932), part 7, pp. 47–50. *The experimental history of colours* (1664), pp. 104, 126–127. The first edition was published in 1663. I am grateful to Marie Boas for this reference.
22. See Florian Cajori, *A History of Physics in its Elementary Branches, Including the Evolution of Physical Laboratories* (New York: Macmillan, 1929), pp. 178 ff.
23. For example, see ibid., p. 184.

10. The Pennsylvania Hospital

Originally published as the introduction to a facsimile reprint of Franklin's *Some Account of the Pennsylvania Hospital* (Baltimore: Johns Hopkins University Press, 1954).

1. It was not included in Albert H. Smyth, ed., *The Writings of Benjamin Franklin* (New York: Macmillan, 1910). Portions are reprinted in *The Papers of Benjamin Franklin,* vol. 5 (New Haven: Yale University Press, 1962).
2. Paul Leicester Ford, *Franklin Bibliography: A List of Books written by, or relating to, Benjamin Franklin* (Brooklyn, N. Y.: privately printed, 1889). The book on the Pennsylvania Hospital is listed on p. 51 as no. 99.
3. Some claim has been made that the Philadelphia General Hospital might antedate the Pennsylvania Hospital. This question is answered in Francis R.

Packard's *Some Account of the Pennsylvania Hospital from its first Rise to the Beginning of the Year 1938* (Philadelphia: Engle Press [for the Pennsylvania Hospital], 1938), a title taken from Franklin's book.

4. See, for example, Carl Van Doren, *Benjamin Franklin* (New York: Viking 1938); Paul Leicester Ford, *The Many-Sided Franklin* (New York: Century, 1899).
5. *Essays upon Several Projects: or, Effectual Ways for Advancing the Interests of the Nation . . .*, in *The Works of Daniel Defoe,* with a memoir of his life and writings, by William Hazlitt (London: John Clements, 1843), vol. 3.
6. Franklin to Samuel Mather, Passy, 12 May 1784, in Smyth, *Writings,* vol. 9, p. 208.
7. See Thomas James Holmes, *Cotton Mather: A Bibliography of his Works* (Cambridge, Mass.: Harvard University Press, 1940), vol. 1, pp. 89–95: *Bonifacius: An Essay upon the Good, that is to be devised and designed by those who desire to answer the great end of life, and to do good while they live . . .* Franklin's youthful satires, directed against the Mathers, were signed "Silence Dogood." Franklin, according to Perry Miller, had "with diabolical cunning" taken "unto himself the cognomen of Silence (since Cotton Mather was possessed, according to the *Courant,* of an irresistible itch for scribbling) Dogood, a blasphemy for which he did penance in his old age" by writing the letter to Samuel Mather about the influence of Cotton's *Essays to Do Good* on his conduct through life. Perry Miller, *The New England Mind: From Colony to Province* (Cambridge, Mass.: Harvard University Press, 1953), p. 410.
8. A new perspective on Mather's scientific interests resulted from three papers by George Lyman Kittredge, "Cotton Mather's Election into the Royal Society," *Publications of the Colonial Society of Massachusetts,* 14 (1913): 81–114; "Further Notes on Cotton Mather and the Royal Society," ibid.: 281–292; "Some Lost Works of Cotton Mather," *Massachusetts Historical Society, Proceedings,* 45 (1911–12): 418–479; "Cotton Mather's Scientific Communications to the Royal Society," *Proceedings of the American Antiquarian Society,* 26 (1916): 18–57.
9. See Theodore Hornberger's essay on Mather's science in Holmes, *Cotton Mather,* vol. 1, pp. 133–137. See also I. Bernard Cohen, ed., *Cotton Mather and American Science and Medicine: With Studies and Documents Concerning the Introduction of Inoculation or Variolation,* 2 vols. (New York: Arno, 1980).
10. Conway Zirkle, *The Beginnings of Plant Hybridization* (Philadelphia: University of Pennsylvania Press, 1935), pp. 103–107; "Gregor Mendel and His Precursors," *Isis,* 42 (1951): 97–104, esp. p. 98.
11. Increase Mather, *Several Reasons proving that Inoculating or Transplanting the Small Pox is a Lawful Practice, and that it has been Blessed by God for the Saving of many a Life;* Cotton Mather, *Sentiments on the Small Pox Inoculated* (Boston, 1721), reprinted with an introduction by George Lyman Kittredge (Cleveland, 1921). The medical aspects of the subject are

reviewed by Reginald H. Fitz, "Zabdiel Boylston, Inoculator, and the Epidemic of Smallpox in Boston in 1721," *Johns Hopkins Hospital Bulletin,* 22 (1911): 315ff. For a view of the whole subject, see Miller, *From Colony to Province,* chap. 21, "The Judgment of the Smallpox."

12. A general study of Mather's medicine is Otho T. Beall, Jr., and Richard M. Shryock, *Cotton Mather, First Significant Figure in American Medicine* (Baltimore: Johns Hopkins University Press, 1954; rept., 1968). This work, which interprets the nature of Mather's medical thought and contributions, includes as an appendix some of the more significant chapters from his *Angel of Bethesda* (1725); see also Cohen, *Cotton Mather.* For a new edition, see *The Angel of Bethesda,* ed. Gordon W. Jones (Barre, Mass.: American Antiquarian Society and Barre Publishers, 1972); see also Margaret Humphreys Warner, "Vindicating the Mininster's Medical Rule: Cotton Mather's Concept of the *Nishmath-Chajim* and the Spiritualization of Medicine," *Journal of the History of Medicine,* 36 (1981): 278–295.
13. Franklin's pamphlet recommending inoculation to Americans is reprinted, with an explanatory introduction, in I. Bernard Cohen, *Benjamin Franklin: His Contribution to the American Tradition* (Indianapolis: Bobbs-Merrill, 1953), pp. 189–199.
14. C. Mather, *Manuductio ad Ministerium: Directions for a Candidate of the Ministry* (1726; rept.; New York: Columbia University Press, 1938); T. J. Holmes: *Cotton Mather* (ref. 7, above), vol. 2, pp. 617–636.
15. Miller, *From Colony to Province,* has shown that Mather referred to Newton as "our perpetual dictator" earlier than the *Manuductio:* The first use of this phrase by Mather was in a "spirtualizing tract" published in 1712, entitled *Thoughts for the Day of Rain.*
16. See Miller, *From Colony to Province,* chap. 24, "Dogood"; *The New England Mind: The Seventeenth Century* (New York: Macmillan 1939), chap. 14, "The Social Covenant."
17. Smyth, *Writings,* vol. 2, pp. 370ff.
18. James Parton, *Life and Times of Benjamin Franklin* (New York: Mason Brothers, 1864), vol. 1, pp. 47–48, 154–162.
19. See Carl Bridenbaugh, *Cities in the Wilderness* (New York: Ronald Press, 1938).
20. See Frederick B. Tolles, *Meeting House and Counting House* (Chapel Hill: University of North Carolina Press, 1948).
21. Franklin's autobiography, in Smyth, *Writings,* vol. 1, p. 352; see Note 9, Chapter 3 above.
22. Ibid., p. 379.
23. Ibid., pp. 376–379. Franklin's own copy (in MS) of his autobiography has been printed *verbatim et literatim* in Max Farrand, ed., *Benjamin Franklin's Memoirs: Parallel Text Edition* (Berkeley: University of California Press, 1949).
24. In the concluding paper in *Meet Dr. Franklin* (Philadelphia: Franklin Institute, 1943), pp. 221–234.
25. P. Clark Sydney, *Pennsylvania Hospital—Since May 11, 1751: Two Hun-*

dred years in Philadelphia (New York: The Newcomen Society in North America, 1951), 44 pp.; Kenneth C. Crain, "Some Account of the Pennsylvania Hospital From its First Rise to the Beginning of the Year 1951," *Hospital Managament,* 71 (1951): 3–23; Harriet C. Crane, "Treasures from the 18th Century Minutes," *The Pennsylvania Hospital Bulletin,* 9 (Winter 1951–52: 1–3, 6–8; E. B. Krumbhaar, "The Pennsylvania Hospital," *Annals of Internal Medicine,* 34 (1951): 1280–1283, and "Days at the Pennsylvania Hospital during its First Century," *The American Journal of Medicine,* 9 (1951): 540–545; Francis R. Packard, "The Practice of Medicine in Philadelphia in the Eighteenth Century," *Annals of Medical History,* 5 (1933): 135–150; Mildred E., Whitcomb, "The Country's First Hospital Puts Service First," *The Modern Hospital* (May 1951): Edmond G. Thomas et al., "Pennsylvania Hospital's 200 years," *Hospitals,* Journal of American Hospital Association (May 1951), and "The Pennsylvania Hospital," *The Pennsylvania Hospital Bulletin,* 9 (1951): 1–23.

Three older publications contain further valuable material. One of these is Packard, *Some Account of the Pennsylvania Hospital to the Year 1938;* another is Thomas G. Morton: *The History of the Pennsylvania Hospital, 1751–1895* (Philadelphia: Times Printing House, 1895); and the third is George B. Wood, *An Address on the Occasion of the Centennial Celebration of the Founding of the Pennsylvania Hospital* (Philadelphia: T. K. and P. G. Collins, 1851).

26. Edward B. Krumbhaar, "The Pennsylvania Hospital," *Transactions of the American Philosophical Society,* 43 (1953): 237–246; this beautifully printed book, profusely illustrated and containing a map showing historic spots in Philadelphia, bears the general title: "Historic Philadelphia from the Founding until the Early 19th Century: Papers Dealing with Its People and Buildings, with an Illustrative Map."
27. See Note 3.
28. See Note 25.
29. See the description in Morton, *History,* pp. 46–49.
30. Morton quotes the following document: "One of the books lately published containing a narrative of the management of the hospital for the last seven years including the account of the last year an abstract of the patients and a list of the contributors having been delivered to the Speaker and by him on Sept. 9th, 1761, communicated to the House of Representatives, William Allen, the Chief Justice and most of the members of the Assembly visited this hospital and after viewing the patients and inspecting the institution were pleased to express themselves much satisfied to observe the decency and economy of the house and that the good purposes of the charity were so carefully attended to." *History,* p. 49.
31. This reprint is described in Ford, *Franklin Bibliography,* p. 51, no. 100. Some extracts have been printed in Cohen, *Benjamin Franklin,* pp. 177–188.
32. See Lawrence C. Wroth, "Benjamin Franklin: The Printer at Work," in

Meet Dr. Franklin (Phildelphia: Franklin Institute, 1943), p. 161; *The Colonial Printer* (Portland, Me.: Southworth-Anthoensen Press, 1938).

33. Van Doren, *Benjamin Franklin,* p. 123.
34. Wroth, "Benjamin Franklin," p. 176.
35. Ibid., p. 175.
36. Dean A. Clark, "Two Centuries of Service—What of the Next?" *The Pennsylvania Hospital Bulletin,* 9 (1951): 2–3.
37. Ibid., p. 4.

11. *The Transit of Mercury*

Originally published as "Benjamin Franklin and the Transit of Mercury," *Proceedings of the American Philosophical Society,* 94 (1950): 222–232. Written with the assistance of a grant from the Penrose Fund of the American Philosophical Society.

1. Jared Sparks, ed., *The Works of Benjamin Franklin,* rev. ed. (Philadelphia: Childs & Peterson, 1840), vol. 6, pp. 159ff.
2. A. H. Smyth, ed., *The Writings of Benjamin Franklin,* vol. 3 (New York: Macmillan, 1907), p. 122.)
3. William J. Campbell, *The Collection of Franklin Imprints in the Museum of the Curtis Publishing Company; with a Short-Title Check List* (Philadelphia: Curtis Publishing Co., 1918), pp. 185–332.
4. C. Evans, *American Bibliography,* vol. 3, *1751, 1764* (Chicago: Blakely Press, 1905).
5. W. C. Ford, *Broadsides, Ballads, etc. Printed in Massachusetts 1639–1800* (Boston: Massachusetts Historical Society, 1922), p. 75; J. Sabin, *Bibliotheca Americana,* ed. R. W. G. Vail (New York, 1936), p. 28.
6. Transcribed from Sparks, *Writings.*
7. This issue of "Poor Richard's Almanack," of great interest because it also contains Franklin's first published account of his newly invented lightning rod, is reproduced in facsimile on pp. 225–260 of Frank Luther Mott and Chester E. Jorgenson, eds., *Benjamin Franklin: Representative Selections* (New York: American Book Company, 1936).
8. *Les membres et les correspondants de l'Académie royale des sciences* (Paris: Académie des sciences de l'institut de France, 1931).
9. *The Letters and Papers of Cadwallader Colden,* vol. 4, Collections of the New-York Historical Society, 1920), pp. 367–368. A version of Alexander's earlier letter to Franklin, dated 26 Jan. 1753, made from a copy, may be found ibid., pp. 363–367. Letters from James Alexander to Franklin relating to the transit are printed, with scholarly annotations, in *The Papers of Benjamin Franklin,* vol. 4 (New Haven: Yale University Press, 1961), pp. 415–422.
10. As in 'sGravesande, *Elements of Natural Philosophy,* 2d. ed., trans. J. T. Desaguliers (London, 1726), vol. 2, p. 155.
11. Mott and Jorgensen, *Representative Selections,* pp. 225ff.

12. For a full account of Winthrop's Harvard expedition to observe the transit of Venus, see I. Bernard Cohen, *Some Early Tools of American Science* (Cambridge, Mass.: Harvard University Press, 1950), chap. 2.
13. Cadwallader Colden Papers, vol. 4, p. 368.
14. Ibid., p. 373.
15. Ibid., p. 382.
16. Ibid., p. 392.
17. Ibid., p. 405.
18. Ibid., pp. 375–376.
19. Ibid., p. 388.
20. Manuscript letter in the American Philosophical Society Library, Franklin Papers, vol. 69, p. 65. My description of its contents is taken from I. Minis Hays, ed., *Calendar of the Papers of Benjamin Franklin in the Library of the American Philosophical Society* (Philadelphia: American Philosophical Society, 1908), vol. 1, p. 6. Hays gives the date as 1752, which is clearly erroneous since the letter refers to the failure to have observed a transit which occurred in 1753.
21. Cadwallader Colden Papers, vol. 4, p. 402.
22. The copy of this letter transcribed in ibid., pp. 393–394, spells the name as "Shewington," whereas the letter as published in the *Philosophical Transactions* spells the name as "Shervington." As to whether Shervington (or Shewington) was merely a bystander or the actual observer, see Note 24.
23. *Philosophical Transactions of the Royal Society,* 48 (1753): 318–319.
24. Note to the first paragraph of the letter: "Dr. Charles Rose, who was in Antigua at this time, says that these observations were taken by Capt. Richard Tyrrel, of the said island, and who is possessed of a valuable collection of astronomical instruments, made by Mr. Bird in the Strand, London, and that Mr. Shervington only was present."

    Note to the last paragraph: "Mr. Shervington has taken the mean of these two altitudes from the error of his watch; and there can be no doubt that his observation is a good one, which, compared with that made in Surry-Street by Mr. Short, p. 199, l. 1, &c., fixes the longitude of the place of his observation in Antigua 4h 5′30″, or 61° 22′30″ west of St. Paul's, London." The reference to "page 199, line 1, &c." signifies an article in the same volume of the *Philosophical Transactions,* pp. 192–200, by Mr. J. Short, on the observations of the transit of Mercury made by him in London.
25. The example of this pamphlet in the Sparks Collection in the Harvard College Library is reproduced in facsimile in *Proceedings of the American Philosophical Society,* 94 (1950): 228–232; it consists of four printed pages and a hand-drawn diagram. The Sparks copy is inscribed by Franklin to "Dr. Johnson," the Samuel Johnson who became president of King's College (now Columbia) in New York. Another copy, in the Massachusetts Historical Society Library, contains the name of James Bowdoin in the latter's hand and is evidently the one sent by Franklin to Bowdoin which is mentioned in Franklin's letter to him, printed in the text of this chapter.

12. Faraday and the "Newborn Baby"

Originally published as "Faraday and Franklin's 'Newborn Baby,' " *Proceedings of the American Philosophical Society,* 131 (1987): 177–182.

1. Seymour L. Chapin, "A Legendary Bon Mot? Franklin's 'What Is the Good of a Newborn Baby?' " *Proceedings of the American Philosophical Society,* 129 (1985): 278–290.
2. See, for example, Richard Gregory, *Discovery, the Spirit of Science* (London, 1916), p. 3. According to Gregory, a lady approached Faraday after one of his lectures on electromagnetism at the Royal Institution, remarking "But, Professor Faraday, even if the effect you explained was obtained, what is the use of it?" Faraday's reply is given as: "Madam, will you tell me the use of a newborn child?"
3. The City Philosophical Society was founded in London in 1808 by John Tatum, the same person whose lectures in natural philosophy Faraday attended in 1810 and 1811, at the very start of his scientific career. On this society, and Faraday's connection with it, see Silvanus P. Thompson, *Michael Faraday, His Life and Work* (London, 1901), pp. 6, 14, 16, 40–41. According to Thompson, "The City Philosophical Society was given up at the time when Mechanics' Institutes were started in London, Tatum selling his apparatus to that established in Fleet Street, the forerunner of the Birkbeck Institution." The texts of Faraday's chemical lectures for the City Philosophical Society are to be found in the library of the Institution of Electrical Engineers, London. On this topic see L. Pearce Williams, *Michael Faraday, a Biography* (London and New York, , 1965), pp. 15–20, 30, 46. Williams, unlike Bence Jones (see Note 4) and Thompson, does not quote Faraday's use of Franklin's image of "a child who is just born" or "an infant," nor does he discuss the Gladstone incident.
4. Quoted in Bence Jones, *Life and Letters of Faraday* (London, 1879), vol. 1, p. 218. See also John Tyndall, *Faraday as a Discoverer* (London, 1868), p. 35.
5. David M. Knight, "Humphry Davy," *Dictionary of Scientific Biography.*
6. Henry Guerlac, *Antoine-Laurent Lavoisier, Chemist and Revolutionary* (New York, 1975), pp. 93, 98, 102, 105.
7. See Knight, "Humphry Davy," pp. 602–603.
8. Alexander Findlay, *A Hundred Years of Chemistry* (New York, 1937), pp. 262–263.
9. Thompson, *Michael Faraday,* pp. 55, 91; L. Pearce Williams, "Michael Faraday," *Dictionary of Scientific Biography.*
10. Quoted in Tyndall, *Faraday,* p. 35.
11. W. E. H. Lecky, *Democracy and Liberty* (London, 1899), vol. 1, p. xxxi. Typical of statements concerning Faraday's two replies about usefulness is the following from James Kendall, *Michael Faraday, Man of Simplicity* (London, 1955), p. 14: "One question they asked him repeatedly, when they visited the Royal Institution and saw him perform some experiment: 'But what's the use of it?' Faraday usually made the same answer that

Benjamin Franklin made under similar circumstance: 'What's the use of a baby? Some day it will grow up!' On one occasion, however, he replied differently. Mr. Gladstone, then Chancellor of the Exchequer, had interrupted him in a description of his work on electricity to put the impatient inquiry: 'But, after all, what use is it?' Like a flash of lightning came the response: 'Why sir, there is every probability that you will soon be able to tax it!' "

12. See I. B. Cohen, "Authenticity of Scientific Anecdotes," *Nature,* 157 (1946): 196–197; reply by R. A. Gregory, ibid., p. 305.
13. Letter to John Pringle, 21 Dec. 1757, in Leonard W. Labaree, ed., *The Papers of Benjamin Franklin* (New Haven, 1963) vol. 7, p. 299. Franklin included this letter in the fourth edition of his book on electricity (London, 1769); he suggested that some "temporary advantage" might result from the patients' "hope of success, enabling them to exert more strength in moving their limbs."
14. On Franklin and mesmerism, see C.-A. Lopez, *Mon Cher Papa: Franklin and the Ladies of Paris* (New Haven, 1966), pp. 168–175.
15. Letter to Peter Collinson, 14 1747, in *Papers,* vol. 3, p. 171.
16. Letter to Sir Joseph Banks, 30 Aug. 1783, in A. H. Smyth, ed., *The Writings of Benjamin Franklin* (New York, 1907–1910), vol. 9, p. 79.
17. See I. B. Cohen, *Franklin and Newton* (Philadelphia, 1956; Cambridge, Mass.: 1966; rev. rept., 1989), chap. 10, "The Franklinian theory of Electricity"; compare John Heilbron, *Electricity in the 17th and 18th Centuries* (Berkeley, 1979), chap. 14, "Benjamin Franklin."
18. See Chapter 6. See also I. B. Cohen, ed., *Benjamin Franklin's Experiments: A New Edition of Franklin's Experiments and Observations in Electricity* (Cambridge, 1942), esp. chap. 3; Note 1, Chapter 1 above.
19. Cohen, *Benjamin Franklin's Experiments,* pp. 100–102; see, further, Heilbron, *Electricity,* pp. 346–349.
20. William Watson, "An account of a treatise, presented to the Royal Society, intituled, 'Letters concerning electricity . . . by the Abbé Nollet . . .' extracted and translated from the French," *Philosophical Transactions,* 48 (1753): 201–202. On the significance of Watson's remarks see Cohen, *Franklin and Newton,* pp. 489–491.
21. Francis Bacon, *Novum organum,* book 1, aphorisms 81, 129; book 2, aphorism 4. Although in the concluding sentence of aphorism 124 (book 1) Bacon stated unequivocally that "truth and utility are here the very same things," he concluded that "works themselves are of greater value as pledges of truth than as contributing to the comforts of life."
22. Carl Van Doren, *Benjamin Franklin* (New York: 1939), p. 700.

Supplement: The Franklin Stove

This essay is dedicated with great affection and gratitude to Charles E. Peterson.

1. *An Account of the New Invented Pennsylvanian Fire-Places: Wherein Their Construction and Manner of Operation is particularly explained . . .*

(Philadelphia: B. Franklin, 1744), reprinted in Leonard W. Labaree, Whitfield Bell, William B. Willcox, et al., eds., *The Papers of Benjamin Franklin* (New Haven: Yale University Press, 1959–1986), vol. 2, pp. 419–446. "Description of a New Stove . . . ," *Transactions of the American Philosophical Society,* 2 (1786): 54–74; reprinted in Albert Henry Smyth, ed., *The Writings of Benjamin Franklin, Collected and Edited with a Life and Introduction* vol. 9, (New York: Macmillan, 1907), pp. 443–462.

2. For more on Franklin's theories of heat, and their relation to the work of Newton and Boerhaave, see Chapter 9, and Note 4 below.
3. Franklin's notion of heat as a "subtle fluid" related to light is summed up in "Loose thoughts on a Universal Fluid," written 25 June 184 and delivered to the American Philosophical Society on 20 June 1788, reprinted in Jared Sparks, ed., *The Works of Benjamin Franklin* . . . (Chicago: Townsend Mac, 1882), vol. 6, pp. 458–460.
4. For Boerhaave's influence on Franklin's understanding of fire and heat, see I. B. Cohen, *Franklin and Newton: An Inquiry into Speculative Newtonian Experimental Science and Franklin's work in Electricity as an Example Thereof* (Philadelphia: American Philosophical Society, 1956; rev. rept., Cambridge, Mass: Harvard University Press, 1989).
5. Nicolas Gauger, *La Mécanique du Feu, ou l'art d'en augmenter les effets, & d'en diminuer la dépense* . . . (Paris, 1713). For more on Gauger's book and its influence on Franklin, see Cohen, *Franklin and Newton,* pp. 261–264. Franklin mistakenly gave the publication date of Gauger's tract as 1709 and misspelled the title; probably, as Cohen believes, Franklin actually consulted the English translation by J. T. Desaguliers (1715).
6. As far as I know, this idea, basic to Franklin's later Pennsylvanian fireplace, is original to Gauger. It survives in the modern "tube grate" and "heatilator" fireplace accessories; see Jay Shelton, *The Woodburners Encyclopedia* (Waitsfield, Vt.: Vermont Crossroads Press, 1976), pp. 83–85.
7. Gauger, *La mécanique du feu,* book 1, part 2, chap. 1.
8. *Pennsylvanian Fire-Place,* pp. 8–10.
9. Vol. 1 was published in 1734, vol. 2 in 1744. See Cohen, *Franklin and Newton,* pp. 243–246.
10. *Fires Improved or a New Method of Building Chimnies* . . . , trans. J. T. Desaguliers (London, 1715 and 1736). The Library Company of Philadelphia copy is inscribed "Gift of Mr. Robert Grace." Franklin owned a second edition (1745) of this treatise; his copy is now in the collection of the American Philosophical Society, Philadelphia. A third "corrected" edition appeared in 1763; there is a copy in the Houghton Library, Harvard University.
11. *Fires Improved,* vol. 2, pp. 555ff, entitled "Postscript; air changed, purified, and conveyed from place to place by the Author."
12. See Cohen, *Franklin and Newton,* pp. 556–558.
13. See Samuel Y. Edgerton, Jr., "Heating Stoves in Eighteenth-Century Philadelphia," *APT* (Bulletin of the Association for Preservation Technology), 3 (1971): 55–104.
14. *Pennsylvanian Fire-Places,* pp. 4–5.

15. Letter to the Marquis de Turgot, 1 May 1781, reprinted in Smyth, *Writings,* vol. 8, pp. 244–245. That Franklin has been thinking of this principle at least since the 1750s is attested in his letter to James Bowdoin, 2 December 1758 (*Papers,* vol. 8, pp. 194–198). He repeated the notion in almost identical words to Lord Kames in a letter dated 28 February 1768, claiming it was a "Law of Nature" (*Papers,* vol. 15, p. 61).
16. For details on the publication of the pamphlet, see *Papers,* vol. 2, pp. 419–421.
17. For Franklin's explanation of this effect, see his letter to Lord Kames cited in Note 15.
18. *Papers,* vol. 2, p. 419.
19. See *Papers,* vol. 2, pp. 419–420.
20. Letter to Roberts, 9 August 1765, in *Papers,* vol. 12, p. 236; letter from Roberts, 27 November 1765, pp. 386–388. The "P Syng" referred to in Roberts's letter as the pattern maker for an otherwise unrecorded version of the Pennsylvanian fireplace was Philip Syng, Jr., a prominent Philadelphia silversmith and cofounder with Franklin of the American Philosophical Society and the Library Company of Philadelphia. Syng was also an early co-experimenter with Franklin on electricity; see I. B. Cohen, *Benjamin Franklin's Experiments* (Cambridge, Mass.: Harvard University Press, 1941), pp. 60–61.
21. "Description of a New Stove," in Smyth, *Writings,* vol. 9, p. 462.
22. Edgerton, "Heating Stoves," p. 23.
23. For instance, Charles Willson Peale wrote in a letter to the *Weekly Magazine of Original Essays,* 31 March 1798: "Formerly [my parlor] was warmed by a fire made in one of the best constructed open stoves being an improvement of Mr. Rittenhouse on Dr. Franklin's stove." Lillian Miller, ed., *The Selected Papers of Charles Willson Peale and His Family* (New Haven: Yale University Press, 1983), vol. 2.1, p. 210. For further documentation of the Rittenhouse stove, see Edgerton, "Heating Stoves," pp. 24–26.
24. See "Letter to Dr. Ingenhousz on the Causes and Cure of Smoky Chimneys," *Transactions of the American Philosophical Society,* 2 (1786), pp. 1–57.
25. In his autobiography and again in his "Description of a New Stove," Franklin accused "an Ironmonger in London" of copying his 1744 pamphlet and making "small changes in the machine, which rather hurt its operation." This ironmonger, Franklin added, received an English patent and earned "a little fortune by it." See Edgerton, "Heating Stoves," p. 50, n. 12; also Josephine Pierce, *Fire on the Hearth* (Springfield, Mass.: Pond-Ekberg, 1951), pp. 42–43, plate 6.
26. Ms. copy in the Library Company of Philadelphia, signed "Hannah Griffiths" and entitled "Inscription on a Curious Stove in the Form of an Urn contrived in such a Manner as to make the Flame descend instead of rising from the Fire, invented by Dr. Franklin," reprinted in Smyth, *Writings,* vol. 1, p. 131.

Index